ii

THE DISCORDANT FORGE

How Johannes Müller, Theodor Meynert, and their rebellious students shaped 19th century scientific thought

Wallace B. Mendelson

Pythagoras Press

ABOUT THE AUTHOR

Wallace Mendelson, MD is Professor of Psychiatry and Clinical Pharmacology (ret.) at the University of Chicago. He is a Distinguished Fellow of the American Psychiatric Association, and a member of the American Academy of Neuropsychopharmacology. He has been director of the Section on Sleep Studies at the National Institute of Mental Health, the Sleep Disorders Center at the Cleveland Clinic Foundation, and the Sleep Research Laboratory at the University of Chicago. His publications include sixteen books and numerous professional papers. He is currently a Fellow of the Faculty of History and Philosophy of Medicine and Pharmacy at the Worshipful Society of Apothecaries, London. Among his honors have been the Academic Achievement Award from the American Sleep Disorders Association in 1999 and a special award for excellence in sleep and psychiatry from the National Sleep Foundation in 2010. More information about Dr. Mendelson's work is available in Wikipedia at https://en.wikipedia.org/wiki/Wallace_B._Mendelson and on his website at http://zhibit.org/wallacemendelson.

CONFLICT OF INTEREST STATEMENT AND DISCLAIMER

CONFLICT OF INTEREST: Dr. Mendelson has no financial arrangements with any pharmaceutical company marketing any medicines mentioned in this book.

DISCLAIMER: This book contains information on a variety of psychiatric disorders as well as their treatments. It is not a substitute for medical evaluation and treatment. If you believe you may have any of the disorders mentioned in this book, please consult your doctor.

TABLE OF CONTENTS

PREFACE

Like so many things in the history of biological psychiatry and psychopharmacology, this book came about due to a serendipitous finding. I had been interested in the life stories of creative individuals who led troubled lives but also made remarkable contributions, often in the face of likely diagnosable mental disorders. These included luminaries in the arts, such as Emily Dickinson and Edgar Allan Poe, and science, such as Charles Darwin (Mendelson, 2021). Most recently I had written on Matthias Jakob Schleiden, who, during the course of a turbulent career in which he twice attempted suicide, revolutionized biology as the co-founder of the principle that cells are the basic building blocks of all living organisms (Mendelson, 2024). At this point I turned my attention to what seemed to be a very different project, describing the historical origins of biological psychiatry, which I had practiced and taught over the course of about forty years. While reading about the emergence of the study of neurophysiology, and the recognition of mental illnesses as disorders of the brain during the 19th century, I came across two remarkable individuals, Johannes Peter Müller and Theodor Meynert.

Müller, under whom Schleiden had trained, and who largely created the field of physiology as a scientific discipline, was reported to have had

1

multiple episodes of depression and died in what was widely interpreted as a suicidal overdose of opiates. Theodor Meynert, who had a remarkable career asserting that specific mental symptoms could be related to brain pathology, was said to have suffered from what was then called 'male hysteria' by no less an authority than his student Sigmund Freud. Indeed, it had been Meynert who had introduced Freud, then a newly-minted physician interested in internal medicine and working in a general hospital, to the field of clinical psychiatry.

Here were two notable figures who had contributed to the genesis of biological psychiatry, whose life stories were entwined with their own possible mental illness. It was indeed a case of serendipity, defined by Horace Walpole, who had coined the term, after noting that "...many excellent discoveries have been made by men who were *a la chasse* of something very different" (Lewis, 1937). I began reading about the lives of these two intriguing historical figures, and to my surprise, I came to realize that in fact their stories were very similar in several intriguing ways:

1. Although they were well known for their original scientific findings, they were equally recognized for the remarkable students they gathered. Müller, for instance, had taught Hermann von Helmholtz, who went on to contribute to the study of energy conservation and the physics involved in vision, Rudolf Virchow, who proposed that diseases could be understood by their effects on cells, and Emil du Bois-Reymond, who founded the field of experimental electrophysiology. Meynert in turn had taught not only Sigmund Freud, but also Carl Wernicke, known for his studies of aphasia and advances in localization of language processing, and Sergei Korsakoff, who discovered the disorder of memory impairment in alcoholics which bears his name.

2. The general consensus was that at some point Müller and Meynert, who had begun their careers with creatively disruptive ideas that altered the trajectory of their fields, may have later become entrenched in their original findings. Thus Müller, it was asserted, failed to take advantage of the new developments in physics and chemistry which advanced physiological research, though he encouraged his students to do so. Similarly, Meynert, it was alleged, had become so focused on his revolutionary emphasis on relating mental symptoms to lesions of specific anatomical areas of the brain that he failed to grasp the importance of disruption of associative brain pathways. It was often asserted that it was their students who then moved beyond the teacher to bring this work to fruition, in the process of creating new disciplines such as clinical electrophysiology, cell biology and histological pathology. Perhaps not surprisingly this conclusion came from narratives written by the students themselves.

3. The third similarity in these histories flows naturally from the second, which is that the relationships of Müller and Meynert with their students were often complex, in some cases involving themes of resentment and alienation on the part of the students. This was most notable in the stories of Müller and Emil du Bois-Reymond, and in Meynert and Sigmund Freud.

In this book I have set out to examine these three themes, which have led me to the conclusion that many advances have been made, not simply from the imparting of wisdom by teachers to students of the next generation, but rather from their very human, and often conflictual, interactions.

REFERENCES FOR THE PREFACE

1. Lewis, W.S. (ed): *Horace Walpole's Correspondence,* Yale University Press, New Haven, 1937, vol. XXXI.

2. Mendelson, W.B.: *The Curious History of Medicines in Psychiatry.* Pythagoras Press, 2020.
    https://www.amazon.com/dp/B083ZRMCW1

INTRODUCTION

A s described in the Preface, this book tells the stories of Johannes Peter Müller and Theodor Meynert, two remarkable 19th century scientists and teachers, emphasizing both their strengths and frailties, and how their legacies were shaped by their interaction with their students. It does not set out to be a complete history of the emergence of modern physiology, which in turn led among other things to a biological understanding of mental illness, but it may be helpful to first briefly describe how their lives fit into that history.

In the late 18th and early 19th centuries, a German worldview known as *Naturphilosophie* arose from the Romantic movement. It viewed nature as a unitary process governed by spiritual forces, and suggested that reason alone is incapable of comprehending nature unless supplemented by artistic and intuitive thinking. This view was captured by Friedrich Schlegel's dictum that "all art should become science and all science art" (Richards, 2002).

Clearly a belief in *Naturphilosophie* was antithetical to the empirical experimentation which was to characterize modern science. In particular, the allied notion of vitalism, which held that all living organisms are governed by an ill-defined vital force, argued against the benefits of mechanistic observation and experimentation. Müller had been heavily influenced by Romanticism in secondary school, as well as when he entered college in Bonn in 1819. As we will see in Chapter One, when he then moved to Berlin in 1823 he found himself admiring the famous anatomist Karl Asmund Rudolphi, who was greatly opposed to Romanticism; a great deal of Müller's subsequent professional life involved the struggle to reconcile these opposing viewpoints, to support his natural inclination toward observation and experimentation. However, in the eyes of some of his students such as Emil du Bois-Reymond (Chapter Two) he never fully moved into the newly emerging fields of chemistry and physics as foundations for understanding living organisms. In their story we see a microcosm of the tumultuous transition from Romanticism to experimental science which was taking place in the mid-19th century.

The growth of physiology as a separate field of study, which took place in those years, also made possible a shift in the way mental illness was understood, resulting in the creation of what was then known as 'brain psychiatry'. Until this time mental illness was often understood as resulting from moral or spiritual failings, imbalances of the four humors described by the ancient Greeks, or malfunction of a hypothetical circulating magnetic fluid. The German psychiatrist and neurologist Wilhelm Griesinger was a leader in suggesting that it was instead the result of disorders of the nervous system. As he phrased it:

"Psychiatry has undergone a transformation in its relation to the rest of medicine. ... This transformation rests principally on the realization that patients with so-called 'mental illnesses' are really individuals with illnesses of the nerves and brain" (Shorter, 1997).

Theodor Meynert gave concrete substance to this approach by emphasizing that mental symptoms could be related to recognizable anatomical lesions of the forebrain, or discrepancies between blood flow to cortical and subcortical structures. In retrospect, many of his specific assertions about lesion sites have proven to be wrong, and his emphasis on single lesions was criticized for not taking into account the notion of disruptions of interconnected networks. His studies, though, were crucial to the growing belief that psychiatry as a discipline was grounded in brain science. The story of Meynert's career in the second half of the 19th century (Chapter Three), then, is a piece of the much broader historical development of the field of biological psychiatry.[1]

I have additionally dwelt at some length on Meynert's interaction with his student Sigmund Freud. This has been done for several reasons: 1. Due to the influence of Meynert and others, Freud, a trained neurologist, had indeed tried to bring together the scientific method and psychology, in his *Project for a Scientific Psychology*; 2. Another influential teacher in Freud's life was Ernst Wilhelm von Brücke, who had been Müller's student, providing a link between the two great teachers who are the subject of this book; and 3. The relation between Meynert and Freud is ripe with analogies to that of Müller and du Bois-Reymond. I find it a fascinating story, especially with the irony that in the 1910s, Freud, now in his 50s, was accused by Carl Jung and Alfred Adler of having gotten stuck at an earlier

---

[1] In passing, it should be noted that this period in the 19th century is sometimes referred to as the 'first wave' of biological psychiatry; it was followed in the latter half of the 20th century by the 'second wave' which grew from the recognition of genetics of mental illness and the development of medicines such as chlorpromazine, lithium and imipramine. The third wave, beginning in the late 20th century and early 21st century, is thought to benefit from the growth of molecular and cognitive neuroscience and advances in neuroimaging (Walter, 2013).

point in his thinking—namely the dominant role of sexuality as the source of the neuroses. The erstwhile rebel now found himself in the uncomfortable position of the master challenged by his former students of being unable to move ahead.

It may also be helpful to provide a brief guide to the overall structure of this book: Chapter One describes the life of Johannes Peter Müller, followed by Chapter Two, which deals with his relationship with his students. Chapter Three presents the life of Theodor Meynert, followed by Chapter Four, which deals with his interactions with Carl Wernicke and Sigmund Freud. Chapter Five compares the stories of Müller and Meynert and their students, noting that although the personalities and tactics involved had many differences, they both also have something to say about how conflicts between teachers and the next generation training in their facilities contributed to the rise of modern science.

Before closing this Introduction, a few words about source materials. I have drawn from Laura Otis's book *Müller's Lab* (2007) as perhaps the most detailed account of Müller's relationship with Emil du Bois-Reymond. Since publication of that work, the teacher's great-grandson Horst Müller published a biography in German (Müller, H., 2019) which includes materials from the family archives and proved to be very helpful. My description of Theodor Meynert's personal life was enriched by his daughter's memories, also published in German almost four decades after his death (Stockert-Meynert, 1930).

REFERENCES FOR THE INTRODUCTION

1. Müller, H.: *Johannes Peter Müller: Physiologe, Anatom, Natur-Forscher*, Books on Demand, Fourth Edition, 2019. ISBN-13: 978-3749489497

2. Richards, R.R.: *The Romantic Conception of Life*. The University of Chicago Press, 2002.

3. Shorter, E. (1997). *A History of Psychiatry: From the Era of the Asylum to the Age of Prozac*. Wiley, 1997, p. 76. ISBN 978-0-471-15749-6.

4. Stockert-Meynert, D.: *Theodor Meynert und Seine Zeit*, Vienna and Leipzig, Austrian Federal Publishing House for Education, Science and Art, 1930.

5. Walter, H.: The third wave of biological psychiatry. Frontiers in Psychology, Vol. 4, Sept. 5, 2013. doi: 10.3389/fpsyg.2013.00582

Chapter One:
Johannes Peter Müller

Müller was a pivotal figure in the development of 19th century physiology. He was known not only for his own studies, but also for his ability to synthesize huge amounts of data into his widely-read handbook of physiology, as well as for his remarkable collection of brilliant students. In addition to those already mentioned in the Preface were Ernst Haeckel, zoologist and proponent of evolutionary theory, Carl Ludwig, who among other things developed a theory of how urine is produced and designed instruments for recording changes in blood pressure across time, as well as Matthias Jakob Schleiden and Theodor Schwann, who proposed that all living organisms were comprised of cells (Mendelson, 2024). Of particular interest among Müller's students was Ernst Wilhelm von Brücke, who later became professor of physiology at the University of Vienna, and in that capacity served as the mentor of Sigmund Freud from 1877 to 1883 (Chapter Four). In these years he grounded his student in the principle that the processes in living things could be understood in terms of chemistry and physics. Interestingly, it was also Brücke who in 1882 convinced Freud to move from his research in neurophysiology to clinical work at Vienna's General Hospital, where he was to later join Meynert's clinic. Let's turn, then, to a summary of Müller's life, and then describe his relationship with his students.

Müller was born in Koblenz, Germany, in 1801, the first of five children of a shoemaker and a coachman's daughter. The family lived in the narrow rooms above the shop. At age seven he was intrigued by the priest's bright robes; he would play a game called 'altar', and thought that someday he might wear them himself. At other times he dreamed of being a knight, displaying his bravery in adventures among the medieval castles along the Rhine (H. Müller, 2019). The view from his window was of a decrepit wall, and among the cracks in the paint he visualized faces and figures which none of his friends could see. Many years later as a scientist he was to look back on this experience when writing his monograph 'On Fantasy Images', and had conversations about it with Johann Wolfgang von Goethe, whose poetic and scientific works he admired.

Müller was a good student and in secondary school excelled at both mathematics and classical languages, reading Plato and Aristotle in the original Greek. His skill was such that he won a prize as the best student of Latin and Greek, sponsored by the director of public education, Josef von Görres, a noted Romantic writer and historian. Johannes was described as dark-haired, with shining eyes; he was robust and athletic, excelling at sports as well as hiking and ice skating. He was an avid collector of insects and plants, early signs of a passion for organizing biological specimens which persisted throughout his life. A peculiarity noted by his fellow students was a strong aversion to spiders. On one occasion he was teased by his classmates for being unable to come into the high school due to fear of a spiderweb near the entrance. He is said to have struggled against this, and in the following year went out of his way to study them.

Upon graduation in 1818, Müller volunteered for one year of military service in the elite 8th Rhenish Pioneer Division in Koblenz. His athletic background aided him in this capacity, though the story was long told that he had once let his mind wander during an inspection; as a consequence, his finger got stuck in his rifle barrel, and, much to the amusement of his friends, it had to be released by the unit's physician.

Müller entered the University of Bonn in 1819. This was a difficult time for him, as his father died, and his mother's efforts to maintain the business did not work out. They mortgaged their home, and he applied for a municipal scholarship, to no avail. Fortunately, in August 1820 the university offered a prize to any student who could "determine through observations and experiments on living animals whether breathing takes place in the fetus as long as it is enclosed in the uterus by the ovum".

When a thorough study of the literature was unable to provide the answer, Müller and his friends searched the countryside on horseback, in order to obtain pregnant animals. After being bitten by a pregnant cat who apparently did not approve of his scientific efforts, he purchased a sheep, in which he found the very different colors of arterial and venous blood in the fetus, indirectly seeming to suggest that a process similar to breathing was taking place. Years later he could not fully confirm his finding in his *Handbook of Physiology*, but it was a remarkable observation for a student project. He was awarded the prize in August 1821. Around this time he was given the Koblenz scholarship which had been denied him earlier, when the prior recipient moved away, and he was also able to make money doing translations from Latin.

This was also an important year for Müller in a personal sense. During Easter church services, he felt deeply touched by the singing of a young lady in the choir. He asked about her, and was told that she was Nanny Zeller, a student at the Koblenz Music Institute, and the daughter of the Koblenz district director. He did not have an opportunity to meet her, however, until he found her at a party in October 1821. They apparently were very taken with each other, and were soon engaged. They were later to marry in 1827 and remained together until his death in 1858.

With his finances now more in order, and his personal life flowering, Müller was able to return to his studies. He obtained his medical degree in Bonn in 1822, with a dissertation on 'Investigations into the laws and numerical relationships of movements in the various animal classes', which included data on the spiders he had collected while a soldier. The finished work also expressed his religious conviction which guided his work for some time. Every living creature, Müller asserted during this early stage of his thinking, had a soul, which could be apprehended through the study of physiology:

"Only the physiologist can understand the soul. Nature never teaches anything different from philosophy. There is no philosophy beyond nature" (doctoral dissertation quoted by H. Müller, 2019).

The search for the elusive 'life force' was the impetus for the many dissections described in his dissertation (H. Müller, 2019). For most of his life he struggled to attempt a reconciliation of vitalism and experimental science, and later, by 1843, he had come to phrase it this way:

"Though there appears to be something in the phenomena of living beings which cannot be explained by ordinary mechanical, physical or chemical laws, much may be so explained, and we may without fear push these explanations as far as we can, so long as we keep to the solid ground of observation and experiment." (Müller, Elements of Physiology, 1843).

At least one author (Meulder, 2010) has suggested that Müller may have shed vitalism toward the end of his career, but that his students du Bois-Reymond, Helmholtz, Brücke and Ludwig continued to portray him as a vitalist. As we will see in Chapter Two, in doing so, they were able to picture him as a man of the previous generation whose beliefs prevented him from advancing into the realms of the more modern physiology which they advocated.

Coming back to our current narrative, Müller graduated with a medical degree at Bonn in December, 1822. His original plan upon graduation was to go to Paris, to study with the famous comparative anatomist Georges Léopold Baron de Cuvier. His application to the minister for funding was approved, with the stipulation that instead he continue his studies in Berlin. After a brief stint as a pathologist in Bonn, he set out for Berlin in May, 1823. Once there, and armed with letters of recommendation from his medical school professors, he circulated among the leading academicians. Karl Asmund Rudolphi, who since 1810 had been the chairman of anatomy and physiology at the University of Berlin, took an interest in him. Rudolphi, who was highly critical of the romantic and philosophical approach to physiology which held sway at the time, disdained the notion of a soul which was responsible for physiologic processes. He gave Müller access to his museum of comparative anatomy as well as his private collection.

After a year and a half of study, Müller returned in October 1824 to Bonn, where he obtained a faculty position and gave regular lectures. His inaugural talk reflected his lifelong struggle to reconcile natural philosophy with Rudolphi's empirical approach: it was entitled 'On the need of physiology for a philosophical view of nature'. As there were minimal facilities available to him, he worked on his collection, which had begun in Berlin, in his apartment, and displayed his specimens in the university's anatomical theater. Within a year, the results were evident. In 1825 he published a paper on egg formation in the ghost locust in the journal of the prestigious Imperial Leopoldine Carolinian Academy of Natural Scientists. In the fall of that year he wrote another paper which secured his fame in the academic community, entitled 'On the comparative physiology of the sense of sight in humans and animals', together with an essay on the movements of the eyes and on the human gaze. He was appointed to a paid position as Secretary of the Academy.

In 1826 Müller was promoted to the rank of Extraordinary Professor (Associate Professor). It was a productive year, in which he began formulating the 'law of specific nerve energies', claiming that the type of sensation produced by stimulating a sensory nerve did not depend on the type of the stimulus, but rather on the characteristics of the sense organ. The retina, for instance, might be stimulated not only by light but also by mechanical means or pressure, but the sensation it evoked was always visual. Around this time he also characterized the functions of the branches of the trigeminal nerve.

In 1826 Müller also gave his fiancé Nanny a paper *Ueber die phantastischen Gesichtserscheinungen* (On Fantasy Images), based in part on his experiences looking out the window of the apartment in his childhood, as we mentioned earlier. In it he described lying in bed with his eyes closed, visualizing people and rooms he had never experienced, before drifting off to sleep, likely what is known today as the normal phenomenon of hypnagogic hallucinations. It concluded that "The sense of sight should soon be examined in its interaction with mental life." Nanny had it published by the young man who had married her sister. She and Müller remained unmarried, largely because of his modest financial situation. In an effort to improve his standing, he briefly tried going into medical practice. He abandoned this effort after a friend died from what was described as an intestinal perforation. Horst Müller commented that 'In any case, Müller's feeling of responsibility was particularly strong and made him anxious and insecure. In addition, there were already enough doctors in Bonn.'

The couple decided that it was now time, about six years since the start of their courtship, to walk down the aisle. About two weeks before the planned ceremony on April 21, 1827, however, Müller began to feel ill. At the time this was attributed to 'the excitement, excessive work and deprivation of the last few years'. He had been following up his interest in visual hallucinations, engaging in sleep deprivation as well as excessive coffee drinking, although the timing of this self-experimentation to the onset of symptoms is not clear. He became irritable, and felt that he was unable to work.

His biographer Horst Müller does not specifically describe feelings of hopelessness or lack of pleasure, though he says Johannes Müller believed he might be dying. The thoughts of death were, however, apparently a source of great anxiety rather than part of thinking that he might be better off dead. He claimed that his legs were paralyzed, and feared that he had a spinal cord disorder. His physician Philipp von Walther diagnosed his condition as hypochondria, reported to the minister that he was already getting better, and suggested that taking a leisurely trip would be therapeutic. The minister agreed with a leave of absence and provided financing; that autumn they left in a rented carriage for southern Germany. He slowly improved, and by winter they were back in Bonn, where he exercised regularly, swam in the Rhine, and returned to work. In February 1828 his daughter was born. He expressed his happiness in letters to his mother, and the family seemed optimistic that things were looking up.

Indeed, things continued to go well. In 1830, at the age of 29, he became the youngest full professor at the University of Bonn. In 1831 his studies in frogs confirmed the work of Charles Bell and Francois Magendie in mammals, indicating that the posterior roots of spinal nerves transmitted sensory information, while the anterior roots were involved with motoric function. In doing so he laid the groundwork for the understanding of spinal reflexes, which were later elaborated by Ivan Pavlov, a generation younger, who studied under Müller's former trainee, Carl Ludwig. It is thought that Müller's emphasis on observable and measurable phenomena, and his experimental approach to physiology, were also indirect but important influences on Pavlov's work (Saugstad, 2018). Müller's assistant in these studies of the spinal cord was the young Theodor Schwann, who later, along with Matthias Jakob Schleiden, proposed that all living things were comprised of cells (Mendelson, 2024).

 The year 1832 was notable for Müller in that he was awarded a gold medal for physiological research by the French Academy of Sciences, and on a personal note, his son Max was born. In November, however, Carl Asmund Rudolphi, his mentor in Berlin, passed away. Müller had always felt close to him, and mourned his loss, but also realized that this meant that the Chairmanship of anatomy and physiology at the University of Berlin was now open. Müller applied in January 1833, and received the appointment in May. He promptly moved to Berlin, and remained in this position until his death.

Within a month, he began a busy lecture schedule, which included general physiology, anatomy of the sensory organs, comparative and pathological anatomy, and reproductive physiology. He brought with him as well the first part of the *Handbuch der Physiologie des Menchen* (*Handbook of the*

Physiology of Men), which was published in German in 1833 and grew to eight volumes when completed in 1840 (later published in English as *Elements of Physiology)*. He was assisted by his trainees including Matthias Jakob Schleiden, mentioned earlier (Mendelson, 2024), and ultimately the project included virtually the entire range of then-known physiology.

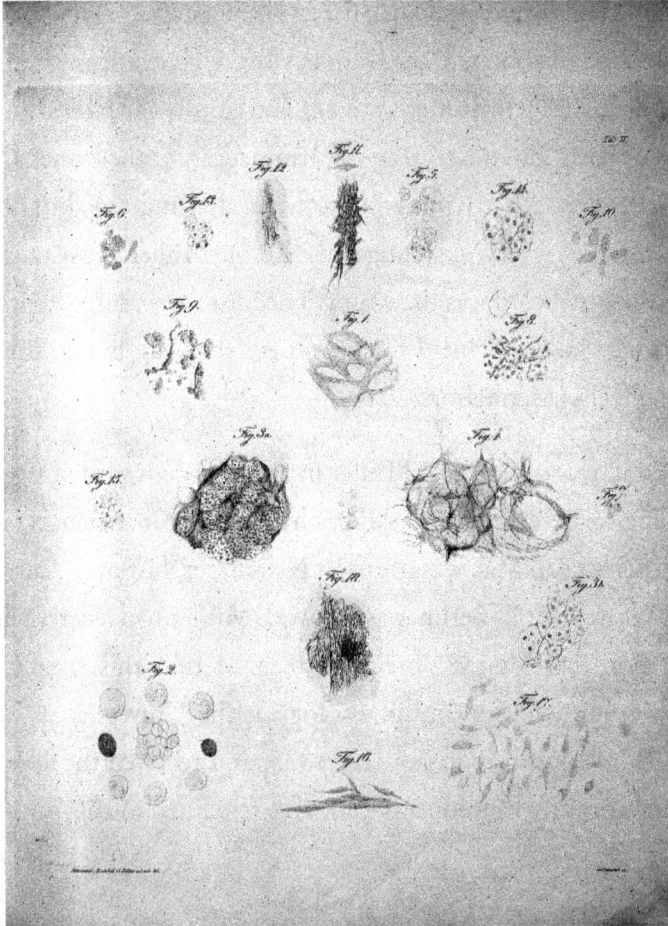

Figure 1-1: *Meynert's illustration of various types of tumors, 1838. Meynert performed some of the earliest microscopic studies of tumor tissue, asserting that they were comprised of cells of many different structural varieties, rather than lymph. In this way he anticipated the work of his student Rudolf Virchow, who is generally credited for establishing the field of cellular pathology.*

Müller could once again spend time in the Berlin Anatomical Museum, and as a sign of commitment he donated his own collection to it. When he took over Rudolphi's position in 1833 the Museum held approximately 7,000 items, and when Müller passed away 25 years later, this had grown to about 19,000. He continued to turn out papers in his own frenetic style. When he made a new finding, he would rush it to publication; in the meantime, he often made additional, related, observations, which he would add to newer, more elaborate papers. As a result, he might publish perhaps four papers on the same topic in a year. He frequently traveled to London and Paris to visit their museum collections and find information not available in Berlin. In 1838 he was appointed to a rectorship[2] at the university and his administrative duties prevented him from traveling so much. He focused on his comparative anatomy studies and the second volume of the *Handbuch*.

Though his work was prospering in many ways in those years, there were some limitations. There was not enough space at the university for his colleagues and students to do their work. In 1835 his trainees Jacob Henle and Theodor Schwann moved to a guest house which Henle described as 'less than a third-rate hotel', and set up equipment for research in their apartments. It was there that Schwann performed the studies that led to his presentation of cell theory, and Henle wrote on anatomy and the germ theory of disease. The university dissection rooms were crowded, and Müller's trainee Du Bois-Reymond likened them to foul-smelling caves (in a similar vein to Auguste-Henri Forel's comments that Meynert's laboratory was disorderly, reminding him of Vienna's Oriental Quarter (Chapter Four). Horst Müller wrote that

---

[2] The highest academic office in universities of the era. The comparable title in contemporary universities in the U.S. might be 'president', and in the U.K. might be 'vice-chancellor'.

"It is easy to imagine that these conditions were unbearable, and yet Müller did not push through the new building at the time. The fact that his wish was constantly denied hurt him deeply. This contributed to the fact that in later times he appeared as inaccessible and dark to the outside world as many people described him" (H. Müller, 1919).

The year 1840 was remarkable for Müller for two accomplishments. He had been searching for some time for a particular shark described by Aristotle, and which had last been seen near Tuscany in 1647. He dispatched his trainee Wilhelm Karl Hartwig Peters to the Mediterranean to retrieve intact specimens of pregnant sharks, and as a result in May 1840 Müller did indeed re-discover Aristotle's shark. Aside from being an anatomical coup, this had special meaning to him, as he had translated Aristotle's theory of dreams in his youth, and remained very much taken with his work throughout his career.

It was in 1840 that Müller also completed the final volume of his *Handbuch*. It was recognized as a remarkable synthesis of the knowledge of the time. Alexander von Humboldt, a linguist, philosopher and government official, and Müller's friend, successfully petitioned King Wilhelm IV to award him the Grand Medal of Art and Science. This work had depended on his ability to bring together vast quantities of scientific data developed by many scholars, but at the same time, he began to feel distress about his personal effectiveness in making contributions. Physiology, which in the early 19th century had a broader meaning, more like our contemporary 'biology' or 'life sciences', was changing; it has sometimes been described as becoming more 'reductionistic and atomistic' (Zeller and Werneburg, 2024). In any event, it had come to be grounded in the new findings in chemistry and physics, which were embraced by his students, but in which he was not trained.

In a sense, then, Müller was in the unhappy position of the Old Testament's Moses, who after leaving Egypt and traveling in the desert for 40 years, was able to see in the distance the new land for his followers, but unable to enter it himself (Deuteronomy 32: 51-52). He believed that he was no longer the leader in experimental physiology, though he realized that he could continue to play a valuable role as a catalyst for his students' work.

This period marked the beginning of a new, and final phase, in Müller's professional life, as he turned from experimentation, and found refuge in morphology and comparative anatomy of marine life, which he pursued with frenetic energy. Whenever his duties as university rector would allow, he traveled to see collections in Sweden, Italy, and France; at other times he would take his younger students on expeditions to collect marine specimens. In 1845, after a trip to Heligoland, a North Sea archipelago, he wrote to Ernst Haeckel: '"Once you enter this magical world of the sea, you will soon see that you cannot escape from it."

During this period, his student Rudolf Virchow (see Figure 1-2) captured both Müller's seemingly default gloomy manner, as well as his ability to transcend it: "What a contrast when the otherwise so gloomy or cold face cleared up with the expression of heartfelt benevolence, when the eyes smiled more than the face and broke through the clouds like a warm ray of sunshine. In such moments Müller was enchanting, for it was precisely then that one became most aware of the man's intellectual greatness" (quoted by H. Müller, 2019).

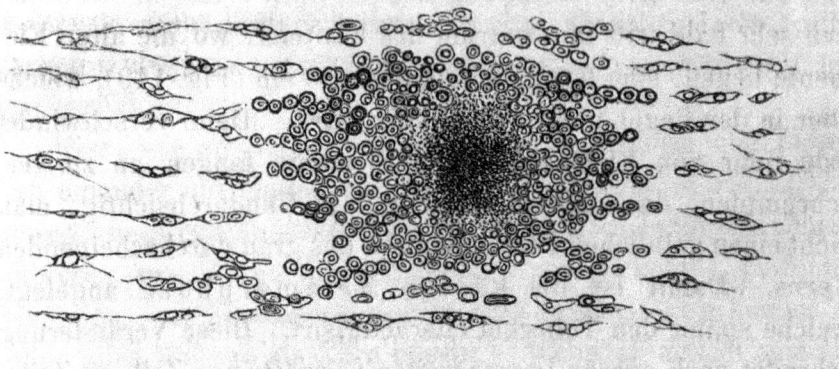

Fig. 140.

Figure 1-2): *Illustration from Rudolf Virchow's lectures at the Pathological Institute of Berlin, 1858. Virchow (1821-1902) applied cell theory to the understanding of pathology, emphasizing that diseases could be best understood by examining their effects on individual cells. He was as well a statesman who believed in dealing with poverty and disease by political action, and advocated universal education. After 1865 he devoted himself to anthropology and archeology, and is credited with establishing anthropology as a scientific field.*

Müller's excitement about his morphological discoveries in the mid-1840s was, however, tainted by two things. The first was the deteriorating condition of the anatomy facilities, which was taking a toll on his trainees. Once again, he petitioned the king for a new building, with the support of Alexander von Humboldt, and once again was turned down. The second was the advent of the revolutions of 1848, which affected him very directly as rector of the university. In March he was faced with student demands, which quickly escalated from the use of lecture halls for meetings, to the abolition of tuition, and eventually to forming armed militias. In all of this, Müller was faced with going between the rebellious students and the Senate, and ultimately the king. The military was brought in, and at some point students were shot. In one incident, the students had dragged an officer into a courtyard where he was being beaten, only to be rescued by Müller alone, who entered the fray in his formal university robes. He was also fearful of damage to his Museum, and guarded it single-handedly day and night while brandishing a cavalry saber.

These events took a significant toll on Müller. In the Fall, as things were becoming more stable, he resigned as rector. He was not sleeping, and was very worried about his financial situation which had deteriorated during this time. He joined his family in Koblenz, but could not find comfort with them. Their conversations, as well as their piano music, irritated hm. Finally, he moved with his son Max to Bonn, where he hoped the quiet life of a scholar would help. He instructed Max in microscopy, but the lessons were inhibited by his irritability and impatience. Ultimately, he turned once again to the sea, moving to Ostend where he did research on echinoderms (a phylum including starfish and sea urchins). Ultimately he began to feel better, and returned to Koblenz. In February 1849 he and Nanny set out on a trip to Marseille where he wished to study, and things began to fall into place. His enthusiasm was so great that he drove long distances himself, rarely taking time to eat. His time in Marseille was successful for him, though Nanny suffered from both his very active pace as well as from being away from the children. In his brief time in Marseille, he was happy to have discovered a new kind of larvae of sea cucumber. In March 1849 he felt able to return to work in Berlin.

Once back in his own facility, he gave a set of keys for the Museum to Alexander von Helmholtz, who had joined him the year before. Helmholtz now had room to work, and was encouraged by Müller to pursue studies of animal electricity, for which he was to become well known (Chapter Two). In 1851 Müller engaged in microscopic studies with his trainee Emil du Bois-Reymond, whose facilities were primarily a hallway for his own students, a room for his research, and in 1854 a lecture room for talks on general physiology. For many years he carried out a great deal of his work at home in his apartment. Later, when du Bois-Reymond succeeded Müller in 1858, he petitioned the Emperor for new facilities, and in 1877 a new physiological institute was opened. His relationship with Müller was complex and will be discussed in the next chapter.

During this time Müller also made trips to Trieste to pursue his study of echinoderms. His personal life in the early 1850s was eventful as well, including the death of his mother, his silver wedding anniversary, and his son Max's graduation. Max planned to become a physician, though Müller had mixed feelings about this, as he had been hoping he would want to become a researcher. By 1854 Müller was flourishing in one sense: he won two outstanding international prizes, the English Copley Medal and the French Prix Cuvier. He was also feeling the weight of his experiences; as early as 1851 at age 50 in a letter to a friend he had said that he was feeling old. He now complained that teaching one course was too much and asked to be relieved of it. He came to feel that social events were excessive obligations; in the past he had willingly danced at parties, and now he was more likely to be seen sitting and watching others. Formerly an enthusiastic concert-goer, he began to fall asleep, to the consternation of his wife.

In 1854 Müller began to describe getting dizzy when looking though the microscope, and sometimes when reaching for higher books on the bookshelves. It was also a year of more political upsets, but he was not directly affected and did not seem excessively disturbed. Then in 1855, while on another collecting expedition, the ship on which he traveled collided with another vessel off the coast of Kristiansand, Norway. He was thrown into the water, but rescued, though one of his students perished.

Upon returning to Berlin he was treated with sympathy, and honored by university officials. He resumed his lectures, where he was greeted by cheers from his students, who presented him with a silver goblet. His wife, Nanny, noted that "The poor little father is so shaken by all these signs of sympathy that his nerves are suffering and he immediately cries." She commented that at home he seemed "softer and more accommodating". Horst Müller (2019) described the change that he was undergoing:

"Now, in Müller's letters, the beginning of a mental breakdown is revealed in a dramatic way, because he feels that the end of his creative powers is approaching. This is also how he interprets the signal of the shipwreck, from which he was only able to escape with difficulty, while the young Wilhelm Schmidt drowned."

The discrepancy between his inner feelings and his outward prosperity continued in March, 1856, when Humboldt informed him that the king commissioned a portrait of him to be hung in the palace collection. In September the family went again to Nice; he continued his research, though it was noted that now he no longer went out in the boats himself, but collected specimens from the fishermen. That winter, Müller had been able to lure his former trainee Rudolf Virchow back to Berlin, and he promptly took over Müller's lectures on physiology and pathological anatomy. Müller had found these progressively harder, as he was no longer involved in research in these areas. His eyesight also seemed to be getting worse, and he often needed a magnifying glass when reading. He again returned to the French Mediterranean coast in 1857 and was seemingly engaged in his marine biology studies. By the winter, though, he apparently had more and more thoughts of death. When parting from a visit with an old friend in

Koblenz, he murmured to him 'We have seen each other for the last time"
(H. Müller, 2019).

Müller's episodes of dizziness continued. He could not sleep, and developed
palpitations as well as a kind of restlessness which led him to wandering by
himself in the streets at night. He complained of abdominal pain for which
he took opium. His teaching load felt oppressive and he longed for time to
work on comparative anatomy, but after dropping the physiology course, his
own studies seemed difficult as well. He became convinced that he could no
longer make contributions, and speculated that his overall approach had no
more to contribute.

On April 28, 1858, he was scheduled to see his doctor, Privy Councillor
Boehm, to make a plan about his declining health. That morning, he
awakened at 5:00 AM and chatted with his wife. Then he seemed to go to
sleep again, and when the family attempted to arouse him two hours later,
he was dead. He had previously stated that when he died there was to be no
autopsy, so the cause of death was never precisely determined. Some
speculated that it may have been from a stroke, but many, including his
trainee Ernst Haeckel, were convinced that it resulted from an overdose of
morphine.

Afterwards, in a moving funeral procession, Haeckel and five more of
Müller's students carried his coffin to the Catholic churchyard in
Luisenstrasse. After the priest presided over the burial, Alexander von
Humboldt stood atop the mound and expressed his love for his old friend.
Later at a ceremony in Müller's former secondary school in Koblenz, an
elderly Rudolf Virchow gave a moving speech in his memory.

REFERENCES FOR CHAPTER ONE

1. Mendelson, W.B.: *From Despair to Discovery: The Botanical Odyssey of Matthias Jakob Schleiden and the Dawn of Cell Theory.* Pythagoras Press, 2024. https://www.amazon.com/dp/B0D3BL2T3Y

2. Muelders, M. (2010). Helmholtz: From enlightenment to

3. neuroscience (p. 962). Cambridge, MA/New York: The

4. MIT Press/Encyclopedia Britannica, Inc.

5. Müller, J.P.: *Elements of Physiology.* Kessinger Publishing, 2010 edition (original translated by William Baly, London 1837-1843).

6. Müller, J.: *Handbuch der Physiologie des Menschen.* Nabu Press, 2013 reprinting.

7. Müller, H.: *Johannes Peter Müller: Physiologe, Anatom, Natur-Forscher.* Books on Demand, 4th edition, 2019.

8. Saugstad, P.: Russian Reflexology. In *A History of Modern Psychology,* Cambridge: Cambridge University Press, pp 149-159, 2018.

9. Zeller, U. and Werneburg, I.: A life dedicated to research and ideal: Johannes Müller between empirical universality and idealistic vitalism mirrored in lecture notes from 1851. Theory Biosci. 143(3):161-182, 2024.

Chapter Two:
Dissention And Discovery--
Müller And His Students

The patrician revolutionary--Emil du Bois-Reymond.

M üller's students described him in a number of ways. This was partly dependent on which portion of his career he was in at the time (that is, before or after he moved from experimental physiology to comparative anatomy in 1840), and varied as well with the background and expectations of the students.

The one that had the greatest influence on how Müller was remembered was probably Emil du Bois-Reymond, who was with him in his later years, in the 1840s and 1850s, and we will focus on him. A fuller account of seven of his remarkable students, notably Ernst Haeckel, who also wrote about Müller, is found in the writings of the American science historian and professor of English, Laura Otis (2007). She has suggested that each drew their own individual picture of Müller, reminiscent of the English author Alethea Hayter's comment on the life of Edgar Allan Poe, saying "Everyone who writes about him chooses their own Poe" (Hayter, 1968). Let's turn, then, to the life of Emil du Bois-Remond, and his interactions with Müller.

<u>Background</u>: Emil du Bois-Reymond was born in Berlin in 1818 to Felix Henri du Bois-Reymond, a Swiss immigrant who became a court counselor for topics related to his native Neuchâtel, and his wife Minette, who came from a well-known Huguenot Berlin family that included scholars, artists and merchants. From this French-speaking home life, he experienced the milieu of intellectual circles in Berlin. He attended the rigorous Französische Gymnasium. Otis (2007) notes that he may have spoken German with a French accent, and at the beginning of the Franco-Prussian war felt awkward because of his French name. These aspects of his background likely left him in the position of an outsider in Berlin. Du Bois-Reymond may have been influenced as well by his parents' Calvinism, but later seemed to lose interest in it. He enrolled in the University of Berlin in 1838. There he was exposed to the *Naturphilosophie* that was prominent in German academic life of the time, but turned to the more mechanistic fields of physics and geology. Later on he was to reject *Naturphilosophie* wholeheartedly, believing that the electrical processes he observed left no room for a mysterious 'life force'.

At this point he settled on studying medicine in 1839, graduating in 1843. His enthusiasm and ability were recognized by Müller, who after meeting him in 1839 brought him into his laboratory in 1840 and soon thereafter entrusted him with the keys to the Anatomical Museum. Müller then guided him into studying the electrical properties of muscle. The issue derived from the earlier studies of Luigi Galvani (1737-1798), who believed that the action of frog nerves and muscle involved electrical processes, in contrast to Allesandro Volta (1745-1827), who thought they did not. More recent studies by Carlo Matteucci (1811-1868) had seemed to strengthen Galvani's view, and Müller suggested that du Bois-Reymond replicate Matteucci's work and further explore its implications. This became the foundation of the major thrust of his life's work. He demonstrated that contracting muscles would cause deflections in galvanometers, devices for sensing electrical current and the precursors of today's ammeters. In other studies he showed that nerve and muscle are stimulated by rapid alterations in current intensity.

Though du Bois-Reymond was inaccurate in the molecular implications of his observations, he consummated the growing belief that nerves and muscles operate by the conduction of electrical impulses. In the process, he ultimately founded the field of experimental electrophysiology. The first edition of his classic *Animal Electricity* was dedicated to Müller, emphasizing that it was his mentor who had guided him in this direction. He was to remain affiliated with Müller, first as an assistant without salary and financially dependent on his father before 1849, and then continuing on until Müller's death in 1858, upon which he succeeded him as professor of physiology.

Figure 2-1: *Close-up of a title page of book by Emil du Bois-Reymond. The story, which apparently originated from an expedition by Alexander von Humboldt to South America in 1800, describes local fishermen who capture electric eels by herding horses into the water; the electric eels then engage*

in a defensive attack in which they shock and disable the horses. Afterwards, the fishermen harvest the exhausted eels. Du Bois-Reymond's interest in electric eels led him to sponsor an expedition to South America by his students to explore the mechanism by which they generate the shocks.

An uneasy alliance-Du Bois-Reymond and Müller: We have already mentioned a number of du Bois-Reymond's thoughts about his working situation in the preceding Chapter. One of his first impressions when he came to work in 1840, according to Horst Müller (2019), was that the dissecting rooms resembled foul-smelling caves. (Referring to the medical students in attendance there, he was also to add in his often-acerbic manner that "the dead company was far less repulsive than the living (du Bois-Reymond, 1840)". As Otis (2007) points out, he came from a well-to-do family on his mother's side; it was difficult for him to be completely dependent on Müller, and to need to ask his assistance for resources. It is not clear whether he appreciated that Müller's own requests for a new building were turned down, and that, according to Horst Müller, this had been painful for Müller. Du Bois-Reymond complained as well that his time was taken up by what he considered to be subservient duties such as working on the anatomical collections in the museum (Finkelstein, 2013).

In spite of these limitations, du Bois-Reymond made rapid strides in electrophysiology over the years. Müller remained his enthusiastic supporter, helping him with promotions in the university system and in gaining general recognition. Müller seemed content in his role as a catalyst of his student's pursuits. In turn, du Bois-Reymond's outward manner was of a loyal assistant, and in the view of his great-great-grandson, who may not have been aware of possibly more complex feelings, "He was the closest to him until Müller's death."

<u>A trainee shapes Müller's legacy:</u> After his death, the manner in which the university chose Müller's successors reflected the breadth of his research interests. A professorship in human and comparative anatomy was given to Karl Bogislaus Reichert (1811-1883), while the professorship in physiology was assigned to du Bois-Reymond. It was in this role that he was often asked about Müller's legacy, and in doing so, he put forth his version, which contained a complex message.

In July 1858, before his appointment as professor, du Bois-Reymond gave a memorial address to the Academy of Sciences, which he later published in a more elaborate form. In Laura Otis's words, "The real aim of the speech is to bury Müller, not to praise him. In this lecture, he tells a story so good that it becomes history" (Otis, 2007). In it he featured Müller's breakdown in 1927 as having been a turning point in which he rejected his earlier philosophical speculations in favor of scientific studies of the frog nervous system. Although he credited Müller with having clarified for the first time Bell and Magendie's thoughts about the differing roles of anterior and posterior spinal nerve roots, he noted that Müller was unable to understand the electrical components of the activity of nerves. The implication was that he, du Bois-Reymond, had moved on to achieve what his mentor had been unable to do.

Du Bois-Reymond's second theme was that Müller's inability to continue on in experimental physiology after 1840 was that in his heart he remained a vitalist, convinced that a vital force inaccessible to the instruments of physiology animated life. For du Bois-Reymond, the latter was crucial in distinguishing Müller's work from his own; he was fascinated with, and devoted to, the technical aspects of instrumentation, which he considered the key to the advancements in physiology. In retreating from the new tools of chemistry and physics, he likened Müller to Satan in Milton's *Paradise*

Lost, who withdrew to what du Bois-Reymond considered a form of Hades, the realm of comparative anatomy (Otis, 2007).

While such colorful allusions and skillful rhetoric—which served du Bois-Reymond well throughout his career—seemed convincing, he was also selective in the memories he chose to present. He made little mention of Müller's scientific legacy, for instance his emphasis on observable, measurable phenomena which his students such as Hermann von Helmholtz and Carl Ludwig continued to teach, inspiring later scientists such as Ivan Pavlov. In the process, du Bois-Reymond created a narrative of Müller's life which he likely hoped would provide a boost to his successful bid to replace his mentor as professor of physiology.

Figure 2-2: *Pendulum device developed by Hermann von Helmholtz for determining the rate at which impulses travel along a nerve. In experiments done around 1850, he used a frog nerve-muscle*

preparation, measuring the time it took from stimulation of the nerve until contraction of the muscle. His 'pendulum myograph' made it possible to accurately measure the short time intervals involved. As a result, in 1850 he determined that nerve impulses in this type of preparation move at about 90 ft/sec. This was remarkably slower than Müller's prediction, about six years earlier, that nerve conduction speed was up to 60 times faster than the speed of light, and indeed was impossible to measure, a view suggestive of supernaturalism compatible with vitalism.

In a multifaceted career von Helmholtz developed mathematical principles of vision, as well as theories of the perception of sound and visual images. He was among the earliest to express a general formulation of the principle of conservation of energy, which ultimately became the first law of thermodynamics.

Other views of Müller:

There are many other examples of how Müller's students shaped his legacy to suit their own purposes. Besides du Bois-Reymond, perhaps the most noticeable one was the zoologist and enthusiast for evolutionary theory, Ernst Haeckel (1834-1919). Otis (2007) argues that, unlike du Bois-Reymond, Haeckel sought to promote himself by emphasizing his similarity to Müller, but that in doing so, he found it necessary to re-create the narrative of his mentor's career. His knowledge of Müller was based on much less contact compared to du Bois-Reymond, and a large part of his descriptions came from his memory several decades after Müller's death (Otis, 2007). Others among Müller's illustrious trainees such as Rudolf Virchow commented on what he was like, but du Bois-Reymond arguably played the biggest role in self-consciously shaping how Müller is remembered. In the next chapter we will turn our attention to the other great mentor who is the subject of this book—Theodor Meynert—describing his life, and in subsequent sections look at his relation with his trainees, and the similarities and differences with Müller's history.

REFERENCES FOR CHAPTER TWO

1. Du Bois-Reymond, E.: *Jugendbriefe*, 42, ketter of 3 February, 1840 (cited in Otis, 2007).

2. Du Bois-Reymond, E.: *On Animal Electricity*. Forgotten Books, 2024 reprint (original 1848).

3. Finkelstein, G.: Emil du Bois-Reymond : Neuroscience, Self and Society in Nineteenth Century Germany. MIT Press, Cambridge, 2013.

4. Hayter, A.: *Opium and the Romantic Imagination*. Faber, 1968.

5. Müller, H.: *Johannes Peter Müller: Physiologe, Anatom, Natur-Forscher*. Books on Demand, 4th edition, 2019.

6. Otis, L.: *Müller's Lab*, Oxford University Press, 2007.

Chapter Three:
Theodor Meynert

Theodor Meynert was born in Dresden in 1833; his father, Günther Meynert, was a journalist and historian. His mother Marie Emmering sang in the opera, where her brilliant voice in a production in Dresden is said to have enchanted Chopin.

When Theodor was eight, the family moved to Vienna. He attended the Piarist Gymnasium, about which his daughter commented:

"Even though my father was inclined towards many arts in his youth, in his family tree, which so many of his colleagues from the Piarist Gymnasium filled with references to their eternal loyalty and prophecies of his future fame, only poetic laurels are mentioned' (Stockert-Meynert, 1930)."

Among his teachers at the Lyceum was Gabriel Seidl, a playwright who encouraged him, and indeed arranged productions of his writings at school. He was encouraged as well by the poet Franz Herz and the playwright and editor Anton Langer.

These were the beginnings of a lifelong interest in poetry, which took many forms. As a schoolboy he was well known for his satirical verses. When a young man; his poems began to appear in literary magazines and later in an Austrian book of poetry (Whitrow, 1996). In his later years he exchanged poetry with Theodor Billroth, who was to become known as the father of modern abdominal surgery.[3] Thirteen years after Meynert's death, his daughter Dora published his collected poems (Meynert, 1905).

In these years he also absorbed the bohemian lifestyle of his parents, which never quite left him. In contrast to this spirit was the influence of his maternal grandfather Andreas Emmerling, who has been portrayed as a strict and duty-bound police physician. He was fascinated by anatomy, and was often involved in studies involving dissection of animals. Because of his position in the police, he was able in 1848 "to intervene and dampen the revolutionary enthusiasm of his then fifteen-year-old grandson Theodor by having him taken into custody without further ado in order to 'prevent him from doing stupid things'". Theodor's daughter Dora further commented that "Papa... often described to us how bitter he felt and how despairing he was when he heard the thunder of cannons outside and knew that those he called his comrades in his heart were in battle" (Stockert-Meynert, 1930).

Meynert, then, was brought up with two competing influences, each of which was to be manifested throughout his life. His grandfather's dedication to duty and to science, and anatomy in particular, was so powerful that much later in 1857, when he died, "In accordance with his wishes, Papa [Meynert, now a physician] dissected him and kept his heart as a specimen for his entire life. When he himself was buried, he took it with him into the ground" (Stockert-Meynert, 1930).

---

[3] Interestingly, Billroth was also an accomplished musician and composer, and was a close friend not only of Meynert, but also Johannes Brahms.

Indeed, it was presumably his grandfather's influence which led him to turn from his bohemian life of poetry to entering the University of Vienna to study medicine. During his student years, however, his unorthodox background remained evident in pranks and hijinks. He was said to sometimes ring all the doorbells on a block, or to exchange their nameplates, or to extinguish the streetlights. Some of these activities landed him in jail, from which he was generally rescued by his grandfather.

Meynert apparently found time as well to study under the pathologist Carl Wedl, who was known for his innovations in histological staining of the myelin sheaths of nerves, as well as his studies on the histopathology of the eye, and on the pathogenesis of various illnesses including tuberculosis and Bright's disease. Additionally he came under the influence of Carl von Rokitansky, a pathologist known among other things for his introduction of integrating information between clinical observation and histopathological findings. Rokitansky, who was also involved in philosophical issues and liberal politics, was to become Meynert's mentor, aiding him in university squabbles later in his career.

Ultimately, after writing a thesis on 'Structure and function of the brain and spinal cord and their significance in disease', Meynert graduated in 1861 at the relatively late age of 28 (Dalvi, 2023). In the time approaching graduation, he had courted Johanna Fleischer, an editor for a fashion magazine, who lived in nearby Klosterneuburg inside the Vienna Woods. They both had a passion for Shakespeare, and one of their favorite activities was for him to read passages to her in the family garden. It was said that his only limitation was pronouncing complicated names, much to her amusement, as she was quite proficient in English, as well as Italian and Latin. She was obviously taken with him, and she must have seemed bold to her more conservative friends, in view of his seemingly meager financial prospects. The two reveled in a pleasant student-like atmosphere, which eventually won over her parents as well. In Meynert's words:

"I became a bridegroom before I was rigorous. If I had been the decent person with an interest in duties and work when Johanna and I were in love, we would either never have become ours or would only have been married after ten years" (Stockert-Meynert, 1930).

Nonetheless, the outcome of their marriage was a new seriousness about his career. Even so, he was to affectionately remember his wild years, and look back at them with nostalgia in older age (Whitrow, 1996).

Meynert chose to forgo developing a private practice, and spoke disparagingly of the possibility:

"What was my calling? To be a doctor. In other words, to compete with the barbers of the seventh district. It would have been necessary to know who lived there, to recommend myself to the priest, to make the acquaintance of some of the area's bigwigs, and many things of the like. All of these things I neglected to do" (Stockert-Meynert, 1930).

Though he had no university appointment, Meynert devoted himself to studying brain tissue, obtained for him by his brother-in-law, who worked in Rokitansky's lab. He was aided by his wife, who contributed anatomical drawings, and his first publications appeared in 1863. He commented at the time that since he was without a formal position, and hence not in competition with his readers, his work was looked upon with indulgence. These first papers, he commented wryly, contained more poetic thought than his formal poetry.

In 1865 Meynert received his first university appointment as Privatdozent[4], allowing him to lecture on neurology and neuropathology, and later became Prosector[5] at the Vienna Asylum. In this position he was able to pursue his studies relating neuropathology to psychiatric illness. In 1868 he was also credentialed in psychiatry. These were good years for him personally as well. He enjoyed spending his evenings at home, reading German classics, as well as Byron and Shakespeare, to his family (Whitrow, 1996). As the children grew older, he often took them to his laboratory, sometimes to the consternation of his employees.

In the early 1870s Meynert's work on the lamination of the cerebral cortex and description of the wide range of cellular types was prospering, but he ran into political troubles when a new director of the Asylum was appointed. The new administration represented the traditional culture of asylum psychiatry, which was far different than the academic orientation of Meynert, who in 1873 had been appointed Professor of Neurology and Psychiatry. The animosity grew to the point that Meynert was asked to leave; happily, his mentor Rokitansky arranged for him to head a newly created second psychiatric clinic at the General Hospital.

Meynert's appointment was controversial in the medical community. It was widely alleged that he had little interest in psychiatry *per se* (a charge which ironically was later leveled against Freud), but over time he prospered, developing a reputation as a researcher and teacher who attracted a variety

---

[4] The rank of Privatdozent does not have a clear counterpart in modern universities, but may be closest to an adjunct professor status. Those who have completed their 'Habilitation', that is, a post-doctoral qualification, are given the right to teach and give examinations to students. It was generally an unpaid position, so that the lecturers received income by collecting fees from their students.

[5] A person who performs autopsies for either clinical or research and teaching purposes.

of promising students. These were to include Carl Wernicke, a German psychiatrist and neuropathologist who was recognized for his work on the localization and processing of language, the Russian psychiatrist Sergei Korsakoff, known for his studies of memory disorders, the Swiss neuroanatomist and psychiatrist Auguste-Henri Forel, and the Austrian neurologist and psychiatrist Gabriel Anton (Chapter Four). Meynert also mentored Josef Breuer, who was to go on to play an important role in Sigmund Freud's growing understanding of hysteria, as well as Freud's close friend Julius Wagner-Jauregg, who was later to become the first psychiatrist to win the Nobel Prize (Mendelson, 1922). Freud himself began working for Meynert in May, 1883. Initially he served as an assistant physician, primarily doing intakes of patients, and then continued neuroanatomical studies for an additional 18 months.

In 1884, Meynert published *Psychiatry: A clinical treatise on diseases of the forebrain based upon a study of its structure, Functions and Nutrition (German 1884, English 1885)*, in which he asserted that mental illness was a manifestation of structural and hemodynamic brain disease, notably disorders of the functional relationship of the cerebral cortex and subcortical zones. In a broader sense, he was trying to have psychiatry recognized as a scientific activity based on brain anatomy and pathology (Seitelberger, 1997). He continued to study the multiplicity of cell types in the cortex, which was later recognized as the beginning of the modern field of cytoarchitectonics. He made advances in cortical localization theory with observations, for instance, about speech defects. Ultimately a number of brain structures were to be named after him, including the substantia innominata of Meynert, the basic optic nucleus of Meynert, and Meynert's decussation. He expanded his clinical work as well, and by 1886 he was making plans to add a neurology clinic, which was to open the following year.

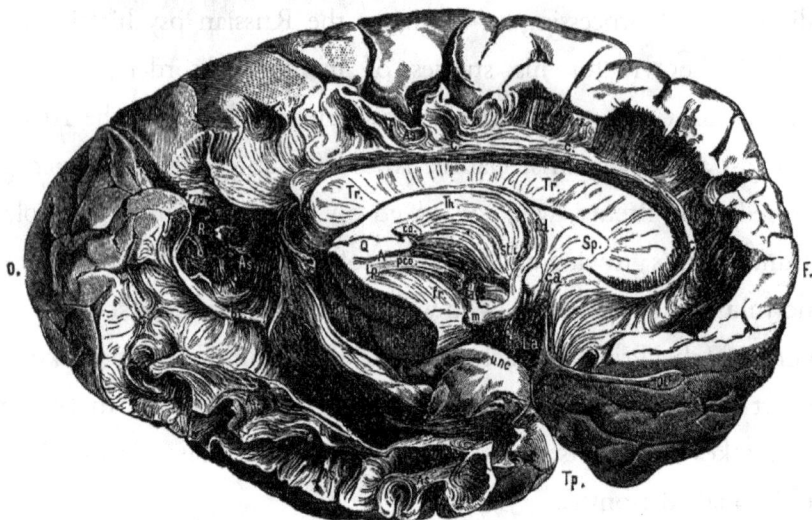

Figure 3-1: *Illustration of dissection of the medial surface of the human brain from Meynert (1885).*

Though he was prospering professionally, these were personally very difficult and complex years for Meynert. In 1878, his mentor and close family friend Carl von Rokitansky died, as did Johanna the following year. In 1882 he remarried Natalie von Grimschit, the daughter of an aristocratic Austrian family, but it was a marriage characterized more by mature companionship than by passion. In 1884, his son Karl developed respiratory disease, and though he was attended by Heinrich von Bamburger (1822-1888), the leading internist in Vienna, he passed away. According to his daughter:

"After her (Johanna's) death and that of his only son, which shattered any hope he had of ever finding this center of gravity again, his only wish remained to be close to their graves. So he settled in Klosterneuburg (the woodsy community

about 10 km from Vienna, in which he had wooed Johanna) where his restless longing and all his happy memories encompassed the Weidlinger Valley" (Stockert-Meynert, 1930).

Thus was Meynert's situation, both professionally and emotionally, by February 1886, when he was 52, at which time his student Sigmund Freud returned from Paris, marking the start of a conflict which was to last until his deathbed some six years later. Before we turn to this conflict which was to change both their lives, let us briefly look at the stories of some of Meynert's other remarkable students.

REFERENCES FOR CHAPTER THREE

1. Dalvi, D. Theodor Meynert (1833-1892)—controversies, contributions and cytoarchitectonics—psychiatry in history. Brit. J. Psychiat. 223(2):388, 2023.

2. Mendelson, W.B.: *The Psychoanalyst and the Nazi Nobelist: The Curious Story of Sigmund Freud and Julius Wagner-Jauregg.* Pythagoras Press, 2022.

3. Meynert, T.: Psychiatry: A clinical treatise on diseases of the forebrain based upon a study of its structure, Functions and Nutrition (translated by B. Sachs), G.P. Putnam's Sons, New York, 1885.

4. Meynert, T.: Gedichte (Wien und Leipzig: Wilhelm Braumüller, 1905).

5. Seitelberger, F.: Theodor Meynert (1833-1892), pioneer and visionary of brain research. J. Hist. Neurosci. 6(3): 264-274, 1997.

6. Whitrow, M.: Theodor Meynert (1833-1892: his life and poetry. *History of Psychiatry* 7(28): 615-628, 1996.

Chapter Four:
Meynert and His Students

M eynert was a complex figure as seen by his students. He was once described as 'a short man conscious of his greatness'. His seemingly large head was crowned with unruly hair which often got in his eyes. His daughter mentioned that his full beard was designed so that he did not look too clerical (Whitrow, 1996). His manner was often ironic and sarcastic. In 1892, Wernicke's student Bernard Sachs (1858-1944) went to Meynert's lab, and is said to have had to work on his anatomical project for a month before getting Meynert's attention:

"A very stormy day," the student commented, attempting to draw him into conversation. Meynert replied "I have not yet had time to think about it", as the frustrated student thought "That settled that" (Medical Eponyms, 2024).

Meynert was described by some of his assistants as trying to be amiable, but usually without warmth, and he did not suffer fools gladly. And underneath his brusqueness was a sense of sadness. Auguste-Henri Forel (1848-1931) in the early 1870s was disappointed in his lectures, and found the laboratory very disorderly, likening it to Vienna's Oriental Quarter.[6] But despite comments about his personal manner, his remarkable ideas drew brilliant students from across Europe.

It is important to recognize that although Meynert was criticized by some of his students for his demeanor, he also evoked long-lasting loyalty among many of them. Notably among these was the German psychiatrist and neuropathologist Carl Wernicke (1848-1905), who achieved substantial recognition as early as age 26 from his work *The Aphasia Symptom-Complex: A Psychological Study on an Anatomic Basis* (1874). A detailed description of Meynert's many students is beyond the scope of this book, but let us look at his relationship with Wernicke in more detail, as it reveals both the strengths—and sometimes weaknesses—of Meynert's approach, and followed a very different trajectory than his relationship with Freud. Those who wish to read a more detailed account of Wernicke's discoveries and influence on understanding forms of encephalopathy, aphasia and mental illness are referred to Ahmad et al. (2024) and Pillman (2007).

---

[6] Forel, a Swiss neuroanatomist and psychiatrist who was also an expert in the social organization of ants, studied with Meynert in 1871, and was not fond of his teaching methods. It was there that he did his early studies of the anatomy of the thalamus. He later went on to characterize the structure of the tegmental region. In 1874, at the age of 26, he published a book on the social organization of ants, *Les Fourmis de la Suisse,* which was praised by Charles Darwin as one of the most interesting books he had ever read. In 1887 he wrote a paper seen as an early formulation of the neuron doctrine (Mendelson, 2023), and in 1889 published a book advocating the role of hypnosis in psychotherapy. After retiring from the directorship of Zurich's Burghölzli Asylum in 1898 and later suffering a stroke in 1911, he wrote on social issues and sexual disorders. Although he received wide public acclaim, including being pictured on both a 1000-frank note and a postage stamp, his reputation began to be re-evaluated in the last few decades, as his writings about eugenics became recognized. He wrote about what he termed 'lower races', speculated about euthanasia of criminals and the mentally ill, and it later came to light that he had practiced sterilization and castration of his patients at the Asylum (Egloff, 2024).

Carl Wernicke

Carl Wernicke was born in May 1848 in Upper Silesia, which was then in Prussia and now in Poland. He went to medical school at the University of Breslau; after graduation in 1870 he remained uncertain as to specialization, and began working as an assistant to ophthalmology professor Ostrid Foerster at Breslau's Allerheiligen Hospital. During the Franco-Prussian War he served as a surgeon in the army, then returned to Breslau as an assistant in psychiatry under Heinrich Neumann. In the early 1870s he was sent by Neumann to work with Meynert for six months. During this period the two developed a close rapport, and in subsequent years Wernicke became very involved with the family, and a frequent visitor in their home. Their closeness was such that it withstood the evolution of his views in neuroanatomy and the basis of mental illness.

The most well-recognized difference from Meynert resulted from Wernicke's 1881 discovery in monkeys of a fiber bundle which he called the vertical occipital fasciculus, or VOF, which may be involved in cognitive and visual processes. Its pathway was vertical, between the dorsal and ventral areas of various parts of the cortex (Yeatman et al., 2014). In 1892, his student Heinrich Sachs (whose unfortunate conversational gambit with Meynert was described previously) confirmed a similar pathway in humans, that had also been reported by Meynert's student Heinrich Obermeyer. For some time, however, the VOF was largely overlooked. The reason is that its existence seemed to contradict a basic principle of organization espoused by Meynert: he believed that all long association bundles ran in an anterior-posterior direction, a position he continued to maintain in a paper written in 1892, the year of his death. In contrast, the VOF, as its name indicated, appeared to run vertically. The VOF did not appear in subsequent brain atlases for some years to come. This may partly be explained by the use of the different names by which it was known by Obermeyer, Wernicke and Sachs, and also a viewpoint questioning its importance, but certainly Meynert's disapproval played a role. It languished for a century, until it was recognized in humans by a diffusion magnetic resonance imaging (dMRI) study (Yeatman et al., 2013).

These strongly differing views, however, did not lead to a parting of the ways of mentor and student. Meynert's daughter Dora noted 'Wernicke's fanatical devotion to Papa' long after he finished his training and became director of psychiatric clinics in Breslau and Halle. She described how he would travel to Vienna to spend evenings with the family. She recalled one occasion in particular:

"Becoming ever more entangled in technical discussions, the two grew further and further apart through the defense of their theses. [The particular subject of this conversation is not clear.] Suddenly Wernicke jumped up and called out to us in a tone of helpless despair: "I am beside myself! This is the first time that I have a different opinion than Meynert" (Stockert-Meynert, 1930).

Throughout, he remained loyal, and attributed his early discoveries in aphasia to Meynert's influence (Whitaker and Etlinger, 1993).

As time went on, Wernicke's views on aphasia and on mental illness grew as well. In general, it can be said that Meynert's conception of mental illness was hierarchical and very anatomical, assigning symptoms to very specific lesions, and focusing attention on differences in blood flow to cortical and sub-cortical structures. In contrast, Wernicke's work on aphasia emphasized functional networks mediated by white matter fiber tracts, and although his theories changed across time, at one point he came to view mental illness as resulting from dysfunction of connective pathways regulating various mental processes including the storage of memory images (Ungvari, 1993)[7].

Some writers have suggested that Wernicke's emphasis on shared connections between functional regions represented a 19th century precursor to the modern interest in neural networks (Ahmad, 2024; Collin, 2016). It is interesting to speculate where his thinking might have led, had his life not been cut short in 1905, at age 57. Meynert's daughter Dora, who would have been in her 30s at the time, reported that news of the bicycle accident on the streets of Halle brought back childhood memories of "one of the most familiar friends in our house for so long".

Gabriel Anton and Sergei Korsakoff

Wernicke's successor as head of the university clinic was Gabriel Anton (1858-1933), the Austrian neurologist/psychiatrist[8] who was another of

---

[7] Interestingly Sigmund Freud, in his 1891 book on aphasia, was highly critical of Wernicke. Labeling him as a 'diagram-maker', Freud believed that it was not necessary to invoke lesions which resulted in disruption of networks, but rather that life experiences might result in pathological disorganization of connectivity. This has been viewed by some as a turning point in Freud's career, at which he began to move away from, or pass through, neuroanatomy.

[8] Anton, who studied under Meynert from 1887 to 1891, went on to make many contributions, including the psychiatric correlates of lesions in the cerebral cortex and basal ganglia, as well as a new approach to the surgical treatment of hydrocephalus. In association with Paul Ferdinand Schilder (1886-1940) he found an association of some forms of chorea to lesions of the lenticular nuclei. He was known as well for his reports, along with Joseph Babinsky (1857-1932), that patients with cortical deafness and blindness may not recognize their deficits.

In his later years Anton gave speeches and wrote materials which were very nationalistic and advocated eugenics and racial hygiene by 'superior breeding'; although he apparently stopped short of clear support for euthanasia, some of his writings could be interpreted in different ways (Kondziella and Frahm-Falkenberg, 2011). His views, then, were not as extreme as those of Auguste Forel, who as we mentioned, was also writing on these topics around the same time, and who openly advocated (and practiced) sterilization and castration of patients with mental disorders.

Meynert's former students, "who not only maintained touching loyalty to him [my father] but also to the entire Meynert family."

The Russian psychiatrist Sergei Korsakoff spent time in Meynert's laboratory, although he was apparently not a formal long-term trainee. His understanding of deficits such as anterograde amnesia and confabulation in the syndrome which bears his name was seen in terms involving dysfunction of memory circuits in the thalamus and mammillary bodies. This, too, represented a movement away from focusing on rigid lesions of specific anatomical locations (Fama et al., 2012). Korsakoff appeared to see this approach as an evolution of thought, rather than criticism of Meynert, and it did not affect their friendship. As we will see in the next part of this chapter, this was not to be the case with the relationship between Meynert and Sigmund Freud.

REFERENCES FOR CARL WERNICKE, GABRIEL ANTON, AND SERGEI KORSAKOFF

1. Ahmad, A. et al.: Carl Wernicke of the Wernicke Area: A historical Review. World Neurosurgery. 185: 225-233, 2024.

2. Collin, G. et al.: Connectomics in schizophrenia: From early pioneers to recent brain network findings. Biological Psychiatry: Cognitive Neuroscience and Neuroimaging. 1(3): 199-208, 2016.

3. Egloff, P.: Auguste Forel (1848-1931) Knocked off his pedestal. Blog of the Swiss National Museum 4/18/24. https://blog.nationalmuseum.ch/en/2024/04/knocked-off-his-pedestal/

4. Fama, R. et al.: Anterograde episodic memory in Korsakoff's syndrome. Neuropsychol. Rev. 22(2): 93-104, 2012.

5. Guenther, K.: The disappearing lesion: Sigmund Freud, sensory-motor physiology, and the beginnings of psychoanalysis. Modern Intellectual History. 10(3): 569-601, 2013.

6. Kondziella, D. and Frahm-Falkenberg, S.: Anton's syndrome and Eugenics. J. Clin. Neurol. 7(2): 96-98, 2011.

7. Medical eponyms (Whonamedit): Theodor Hermann Meynert. https://www.whonamedit.com/doctor.cfm/958.html?t&utm

8. Mendelson, W.B.: The Battle Over the Butterflies of the Soul: Santiago Ramón y Cajal and the Birth of Neuroscience. Pythagoras Press, 2023. https://kdp.amazon.com/amazon-dp-action/us/dualbookshelf.marketplacelink/B0BRJNYNTB

9. Pillman, F.: Carl Wernicke and the neurobiological paradigm in psychiatry. Acta Neuropsychologica 5(4): 246-260, 2007.

10. Ungvari, G.S.: The Wernicke-Kleist-Leonhard School of Psychiatry. Biol. Psychiat. 34: 749-752, 1993.

11. Wernicke, Carl. *Der aphasische Symptomencomplex: eine psychologische Studie auf anatomischer Basis*. (The Aphasia Symptom-Complex: A Psychological Study on an Anatomic Basis). Cohn & Weigert, 1874.

12. Whitaker, H.A. and Etlinger, S.C.: Theodor Meynert's contribution to classical 19th century aphasia. Brain Lang. 45: 560-571, 1993.

13. Yeatman, JD et al.: Anatomy of the visual word form area: Adjacent cortical circuits and long-range white matter connections. *Brain Lang* **125**, 146–155, 2013.

14. Yeatman, J.D.: The vertical occipital fasciculus: A century of controversy resolved by in vivo measurements, *Proc. Natl. Acad. Sci. U.S.A.* 111 (48) E5214-E5223, 2014.

Sigmund Freud and his relationship with Meynert

Freud's background: Before exploring the interaction of Freud and Meynert, it may be helpful to briefly review Freud's early years and the influences of his teachers. He was born on May 6, 1856 in Freiburg, Moravia, to parents of Ashkenazi Jewish descent (Mendelson, 2022). Jakob Freud, his father, was a twice-widowed wool merchant. The family's income was modest, and they lived in rented space in a locksmith's house. Sigismund (later Sigmund) was the first child of the 20-year-old Amalia Freud, who referred to him as her 'golden Sigi'. He was to remember their closeness as something approaching unconditional love, just as he tended to retain thoughts of his father's jealousy. It has sometimes been speculated that decades later, in the 1890s, these memories may have contributed to his formulation of the Oedipal syndrome.

In 1859, the family moved for financial reasons, and after a brief sojourn in Leipzig, came to Vienna. The four-year-old Sigmund was to find the urban life there very different from that in the rural, forested hills of Moravia. They lived in the lower class, largely Jewish neighborhood of Leopoldstadt. In this setting, he came to feel like an outsider in predominantly Catholic Vienna, where antisemitic feelings were not uncommon. This may have led him to feel drawn to stories of Hannibal, the Semitic third century BC Carthaginian general, whose father made him pledge to destroy the Romans (Crews, 2017). He similarly took great interest in Napoleon, Cesar and Alexander.

Interestingly, although Freud and Meynert both moved to Vienna in their childhood, and thus might seem to have similar backgrounds, they actually experienced Vienna in very different ways. As we have described, Meynert was brought up as part of an artistic, bohemian community, which was very tolerant of hijinks and outré behavior, in which he apparently felt at home. Freud, with his Jewish background and financially insecure family, living in a largely Jewish neighborhood, was, as we have described, very much the outsider.

Whatever misgivings Freud may have had about his place in the Viennese community, he prospered academically, and spoke five languages by the time he finished secondary school. At the time he thought of becoming a lawyer, and having a career in politics. When he entered the University of Vienna in 1873, at age 17, his interests had changed to science and medicine. It has been speculated that his decision was possibly influenced by his readings of Goethe's poetry and works on nature. Though he ultimately turned to medicine, he also attended classes in zoology and philosophy, which may have contributed to his relatively late graduation. At one point he spent a month in Trieste, searching unsuccessfully for mature male sex organs in the eel, which was later to be a source of amusement to his psychoanalytic followers.

Influence of Ernst Wilhelm von Brücke and Josef Breuer: Among his teachers, two particularly stood out. His biographer Ernest Jones concluded that Theodor Meynert's lectures interested him more than any others (Jones, 1956). The second was Ernst Wilhelm von Brücke (Chapter One), under whom he began to study and did so intermittently, along with work until 1882 at a pathology institute founded by Brücke's former student Salomon Stricker. Because of Brücke's important influence on Freud, it may be well to briefly describe his career.

Brücke had been Muller's student starting in 1843, going on to teach anatomy in Berlin and physiology in Königsberg, coming to Vienna in 1849. By the 1870s he was the head of his own Institute of Physiology, and was well known for his belief that the principles of natural science could be applied to human thought and behavior. In his lab Freud studied, among other things, the anatomy of the eye and the internal components of nerve cells. When he graduated in 1881, he had already published six papers on various topics in neuroanatomy.

Figure 4-1: *Freud's microtome used for preparing slices of brain tissue for microscopic study. In the late 1770s and early 1880s, Freud worked in the laboratory of Ernst von Brücke (who in turn had previously been a student of Johannes Müller), and in 1885 received an appointment as Privatdozent, or private lecturer, in neuropathology (please see footnote describing this rank in Chapter Three regarding Theodor Meynert). His work is thought to have supported the 'neuron doctrine'*, and was cited by Santiago Ramón y Cajal**.*

Although Freud was later, of course, to turn his attention to the development of psychoanalysis, this microtome is a good reminder of his thorough background in neurobiology. It is believed to have been given to him by his colleague Dr. Carl Koller, and is on display at the Sigmund Freud Museum in Vienna.

*Mendelson, W.B.: The Battle over the Butterflies of the Soul: Camillo Golgi, Santiago Ramón y Cajal and the Birth of Neuroscience. *Pythagoras Press, 2023. **Triarhou, L.C.: Exploring the mind with a microscope: Freud's beginnings in neurobiology. Hellenic J. Psychology 6:1-13, 2009.* https://www.academia.edu/9914521/Exploring_the_mind_with_a_microscope_Freuds_beginnings_in_Neurobiology

In Brücke's laboratory Freud also met Josef Breuer, who had an extensive background in studies of the role of the inner ear in balance, as well as a respiratory reflex which bore his name (the Hering-Breuer reflex). He was now in private practice, but retaining his interest in research. Later he, like Freud, was to join Meynert's clinic.

Breuer's influence on Freud was significant in a number of ways. Among them, he described in detail his patient Bertha Pappenheim whom he had treated for hysteria, initially with hypnosis, from December 1880 to June 1882, which may have stimulated Freud's interest in the disorder; indeed later in 1895 they were to jointly publish *Studies in Hysteria*, in which she was presented under the pseudonym of 'Anna O'. Freud came to see Breuer, who was 14 years older, as a father figure. He was very close to Breuer's family, and later named his first daughter after Breuer's wife.

By 1882, Freud began to consider a career in internal medicine, but his training application was turned down by the two prominent Viennese clinics. Although his research was going well, he was doing very poorly financially, and Brücke recommended that he turn to clinical medicine. At his suggestion, Freud began work at Vienna's General Hospital, where he assisted in a variety of settings. He saw a number of aphasic patients, and later drew upon his experiences in a book on aphasia in 1891. His temporary assignment in an asylum may have also increased his interest in clinical care. Ultimately, he came to work in Theodor Meynert's Second Psychiatric Clinic.

Meynert's influence in the early years: Initially Freud worked as an assistant physician, primarily doing intakes. In retrospect, some who have reviewed Freud's case notes described them as vague and imprecise, which has sometimes been interpreted as lack of engagement on Freud's part. In contrast, others have emphasized his detailed accounts of psychotic disorders including chronic paranoia, as well as dementia paralytica (Mendelson, 2022), indicating his interest and involvement in major psychiatric conditions (Dalzell, 2011). During this period he also sharpened his skills in localizing lesions in relation to clinical symptoms in more clearly organic disorders.

As time went on, Freud turned his attention more to Meynert's neuroanatomic studies. He came to see Meynert as a mentor "in whose footsteps I had trodden with such deep veneration" (Freud, 1900). In these years he was exposed to Meynert's emphasis on the neurological underpinnings of psychiatry, and his belief that psychiatric illness could be associated with lesions in specific anatomical areas. This approach led to Freud's early attempts to see psychic processes as a result of disordered neurobiology, as expressed in his *Project for a Scientific Psychology*. Freud was no doubt influenced as well by Meynert's criticism of unscientific theories of mental illness, which may have led him to attempt to avoid these pitfalls, such as facile notions of heredity (Dalzell, 2011).

Interestingly Meynert also developed an early concept of the ego, based on an individual's unique pattern of brain connectivity resulting from thoughts and experiences (Peled, 2019). It is likely that Freud was influenced in this thinking when he later formed a more abstract view of the ego as part of his structural model of the mind. Meynert had expressed thoughts on repression, defense and the pleasure principle (Danzell, 2011), all of which were to influence Freud in the later development of psychoanalysis. Freud of course did not casually accept such notions, but was to go on to elaborate on them based on clinical observation and to systematize them into a coherent body of thought in psychoanalysis.

Freud later began to question the clinical inferences Meynert drew from his anatomical observations. While Meynert, for instance, described what was then known as amentia (hallucinatory confusion) as being due to compromised blood supply, Freud came to view it as the result of a defense causing a loss of recognition of external reality. By 1891, he was decrying what he considered to be a simplistic association of psychological phenomena to precise anatomical locations as being a form of brain mythology (Freud, 1891). Meynert, for instance wrote that 'Fear is the neurosis of the oblongata, excitation which defies cortical inhibition' (Meynert, 2020 edition).

In making such claims, Meynert was attempting to give anatomical precision to the growing viewpoint espoused by the German neurologist and psychiatrist Wilhelm Griesinger (1817-1868) that mental illness should be viewed as representing disorders of the brain. Freud was not alone in his criticism of such a strictly anatomical approach; the Swiss-German psychiatrist Karl Jaspers (1883-1969) was later to become one of the leaders advocating the view that it was reductionistic to consider all mental disorders to be due to localized brain pathology. In his view, brain processes are necessary for mental activity, but are not in themselves sufficient for describing subjective experiences, which in turn require a more humanistic approach. In summary, Meynert was criticized on the one hand by his fellow physiologists for being limited by focusing on specific anatomical sites rather than disruption of networks, and on the other hand by those who believed that human experience could contribute to brain dysfunction, and that brain pathology alone is not adequate to capture the nature of subjective states.

In 1885 Freud became a 'Privatdozent', allowing him to lecture on neuropathology in an unpaid position at the University of Vienna. He looked into financial support from the university, and received a travel grant to work in Paris under Jean-Martin Charcot. During this period, Meynert clearly thought highly of him; his daughter Dora referred to 'Dr. Siegmund Freud, the world-famous dream psychologist, whose genius my father happily recognized, even though it diverged from his own path (Stockert-Meynert, 1930). Freud, however, saw this new opportunity as a big step. He had the option to take a six-month period of leave from the General Hospital, but instead he chose to resign.

Jean-Martin Charcot: In Paris, Freud became intrigued with Charcot's presentations of patients with hysteria, which was thought to be due to 'ungovernable emotional excess'. At least initially this was found primarily in women, whose symptoms included paralyses, tics, difficulties speaking and other phenomena for which the cause was unclear. Charcot claimed that these could be cured by hypnosis, and seemed to show that hypnosis could produce similar symptoms in non-patients. Some critics considered his dramatic demonstrations to border on the theatrical, and some of his claims have been contested, but at the time they carried significance. They were, after all, coming from a famous neurologist who had founded one of Europe's leading neurological centers at the Salpêtrière Hospital, and who was known for his studies of many diseases including Parkinson's, amyotrophic lateral sclerosis, and multiple sclerosis. Charcot's work on hysteria seemed to suggest that there were mental processes which were not conscious, which of course were later to become a crucial aspect of Freud's topographic conception of mental functioning. Charcot also suggested that patients with hysteria usually had histories of (often sexual) trauma, which again formed part of Freud's early thinking.

Freud's return to Vienna and conflict with Meynert: After arriving back in Vienna in February 1886, Freud was eager to acquaint his Viennese colleagues with his experiences in Paris. He began with lectures in May at The Physiological Club and the Psychiatric Society, but was met with minimal enthusiasm. In October he talked on the topic of male hysteria to the Medical Society. His friend and colleague Julius Wagner-Jauregg, in describing the response, wrote that the president of the society as well as Meynert were very critical of the presentation:

"[Heinrich von] Bamberger and Meynert in the discussion bluntly rejected Freud's statements and thus he fell into disgrace with the Faculty" (Wagner-Jauregg, 1950).

Among other things, Meynert challenged Freud to demonstrate that male hysterics actually exist. Although Freud did indeed do so in November (Dalzell, 2011), Meynert and other luminaries remained unconvinced. Meynert withdrew his earlier offer that Freud might have access to his facilities, leaving Freud with nowhere to lecture (Whitrow, 1990). Freud followed the advice of Josef Breuer, and in the spring set up a private practice which was to continue until 1938. He was very much aware, though, of the difficulty of growing his practice in the absence of support from the leaders of the Viennese academic psychiatric community.[9]

---

[9] A related issue, though not central to the thrust of this book, is whether Freud was even considered by his colleagues to be a psychiatrist. Over the decades, much has been made of the observation that although in 1885 Freud qualified in neuropathology, he never subsequently applied for a Dozentur in psychiatry. As a consequence, to this day, many in the psychoanalytic community believe that he should not be considered a psychiatrist, while others point to his period of training inside Meynert's psychiatric clinic.

Whitrow (1990) argues that, had Freud applied for accreditation in psychiatry, he would likely have been unsuccessful. Most of his clinical publications at that time dealt with

The skepticism and outright hostility to the issue of male hysteria extended as well to Charcot's use of hypnosis. By June 1888, Meynert lectured on hypnosis, referring to it as a 'beastly subjugation' of one person by another (Dalzell, 2011). He speculated that any possible benefits might result from alterations in blood supply, but called it vulgar and the work of charlatans. At a Medical Society lecture in 1889, he attacked not only Charcot, but also Freud, in a very personal way, denying his skill as a doctor and calling him "a skilled practitioner of hypnosis back from Paris" (Dalzell, 2011). In Freud's words, Meynert, 'after a short period of favour, had turned to undisguised hostility' (Freud, 1900).

Freud's reconciliation with Meynert: As time went on, the differences in Freud and Meynert's understanding of hysteria became clearer. On a conceptual level, Meynert declared that hysterical paralysis resulted from alterations in blood flow to the choroidal artery, while Freud elaborated on Charcot's concept of a 'functional dynamic lesion' (Dalvi, 2023). Initially, Freud's neurological practice did not flourish, and in what some believe was partly an economic decision (Fancher, 1979), he let it be known that he was available for treating patients with hysteria. We can only speculate about how he might have felt about this: hysteria had not been a central interest to him when he first departed Vienna to study with Charcot, but when he

---

neuroses, which were considered far from mainstream psychiatry, and were only minimally mentioned in the contemporary textbooks. In 1899, for instance, Julius Wagner-Jauregg wrote in an evaluation for promotion that "Dr. Freud is only Dozent for neuropathology and has never really worked in psychiatry." We can only speculate whether Freud surmised that disapproval by Meynert and others would make such a bid unsuccessful. His not applying was perhaps further evidence of his growing alienation from the institutional psychiatric establishment.

returned, his embrace of Charcot's ideas had led to his downfall in academic circles. Now he was hoping for treatment of hysteria to save his practice.

Since few physicians were willing to accept patients with possible hysteria, Freud received many referrals. In dealing with this influx of cases, however, he began to recognize the limitations of hypnosis in eliciting cathartic emotions and memories, and began to look for other approaches. First, he tried a 'pressure technique' in which the patient lay on a couch with eyes closed, and he would place his hand on their forehead and ask them to recall the evolution of their symptoms. Later he abandoned the hand pressure, and instead would ask the patient to say whatever came to mind, ultimately giving birth to the technique of free association. Analyzing the associations and memories which came forth, as well as the patient's relationship with the therapist, became cornerstones of what ultimately grew into the psychoanalytic movement.

In these years, both Freud and Meynert prospered, each in their own way, and at very different points in their lives. Freud's model of the private practice physician-scientist, which he had adapted from Breuer, turned out to fit him well. In this endeavor he was aided by his clinical skill, his ideas, and an engaging manner with his trainees as the urbane professor. In contrast, Meynert continued to grow inside the classical university setting. In terms of his manner with colleagues and students, he seemed to try to be pleasant, but he was often seen as distant, and preoccupied with his own thoughts. Those who were less fond of him thought of him as a poor teacher, though generally acknowledging the brilliance of his ideas.

When Meynert was approaching death in 1892, Freud visited him, and they apparently had a reconciliation. There are many comments in the literature that Meynert confessed that one of the reasons he had been so hostile to the notion of male hysteria was that he had been trying to hide the fact that he, himself, suffered from it (e.g., Francher, 1996). Such a claim, which brings a narrative of their relationship full circle, has a certain mythic quality to it, but indeed is derived directly from Freud. In *The Interpretation of Dreams*, in the context of analyzing a dream in which he believed that his father represented Meynert, he commented about this deathbed visit:

"When I visited him during his fatal illness and asked after his condition, he spoke at some length about his state and ended with these words: 'You know, I was always one of the clearest cases of male hysteria.' He was thus admitting, to my satisfaction and astonishment, what he had for so long obstinately contested" (Freud, 1900).

It is difficult to assess Freud's recounting of this visit. This was an issue of great emotion as well as practical significance for his career. The hostile reception he received from Meynert following his October 1886 lecture on male hysteria 'rankled with Freud for the rest of his life' (Whitrow, 1990). He was also writing about it eight years after the event, in the context of analyzing a dream. There is also the question of whether, if Meynert was truly 'one of the clearest cases of male hysteria', why this had apparently never been noticed by others. This would be truly remarkable, given the often-dramatic nature of the symptoms, which could involve, for instance, paralyses, fainting, or inability to speak. It remains possible, though, that Meynert had some *forme fruste* of the condition.

It is also uncertain whether, on his deathbed, Meynert was trying to make peace with his former protégé, and of course his own mental clarity at that moment is unclear. In February 1892 his daughter Christi had become ill; she was operated on in their home by Theodor Billroth, his friend and perhaps the foremost surgeon in Germany, but he had been unable to save her. Certainly grief over her death, and the earlier deaths of Johanna and their son Karl, would have been weighing heavily on him during this time. On the other hand, a few days before the end, he was able to present "New Studies on the Association Bundles of the Hiru Mantle" to the Academy of Sciences (Stockert-Meynert, 1930). There were so many factors in play that we may never know what happened during Freud's visit, though it seems likely that some sort of reconciliation took place. Ironically, it came about as a result of both men changing their views: Freud had abandoned hypnosis, while Meynert was seemingly recognizing hysteria in males.

We know little about Meynert's last days in May 1892. His daughter mentions that "just before he fell into agony, [he] sang the... barcarolle from "The Mute Girl of Portici" (Stockert-Meynert, 1930). As this opera has often been associated with revolutionary activity (Slatin, 1979), she interpreted it to mean that "something revolutionary remained in him long after his youth". We know as well that earlier, in 'his declining years' he had written a poem, *Gozian* (his nickname) in which he looked back fondly at the playfulness and hijinks of his undergraduate days (Whitrow, 1996).

As mentioned earlier, Meynert had been opposed to going into private practice as a young man, and throughout his career disagreed with the idea of clinicians practicing their profession as a kind of business, a view for which he was often criticized by colleagues. As his daughter relates, it must have weighed on him on his deathbed that his idealistic view had had an economic effect on his family:

"Shortly before his death, he, who could never be satisfied with his concern for his loved ones, became aware that he had damaged our worldly possessions through the steadfastness of his convictions, and he felt compelled to formally

apologize to me and my sister for this. As if he had not given our existence the greatest gift by being like this!" (Stockert-Meynert, 1930).

We've also seen that in his last two years, he became very close with Theodor Billroth, though they had known each other for decades. This new development in their relationship began when Meynert sent him a poem in January 1891, and they began exchanging verses. In May, Billroth was not able to visit him because of his own ill health. After Meynert died unexpectedly, Billroth captured his complex personality, and the contrast between his outward manner and inward life, in this letter to his wife:

"Only late on did coincidences and the greatest reverence for the deceased's eminent intellectual products bring me closer to him and allow me to discover in him a depth of soul that was often hidden under acerbic humor and was recognized by few! I consider it fortunate that I was still granted the opportunity to take a few glimpses into his tender emotional life!" (Stockert-Meynert, 1930).

Billroth also commented that he would likely be the next to die among their close academic friends, and in this he was correct.

This chapter concludes the portions of this book focusing on the lives of Johannes Müller and Theodor Meynert, and their interaction with their students. In the following chapter we will address the broader picture, looking at their similarities and differences, and their role in the history of science.

REFERENCES FOR SIGMUND FREUD AND HIS RELATIONSHIP TO MEYNERT

1. Crews, F.: *Freud: the Making of an Illusion,* 2017, p. 11.

2. Dalvi, D. Theodor Meynert (1833-1892)—controversies, contributions and cytoarchitectonics—psychiatry in history. Brit. J. Psychiat. 223(2):388, 2023.

3. Dalzell, T.: What Freud learned in Theodor Meynert's Clinic. The Irish Journal for Lacanian Psychoanalysis. 49: 65-72, 2011.

4. Fancher, R.E.: Pioneers of Psychology. London, W.W. Norton, 1979, p. 372

5. Freud, S.: Zur Auffassung der Aphasie. Leipzin, Wien : Braumüller, 1891.

6. Freud, S.: *The Interpretation of Dreams,*1900. In James Strachey's translation of Volumes IV and V of the 1953 *Standard Edition* (London, Hogarth Press and the Institute of Psychoanalysis), Basic Books, New York, pp. 527-528.

7. Jones, E.: *The life and work of Sigmund Freud. I. The young Freud (1856-1900).* London, Hogarth, 1956, p. 72.

8. Mendelson, W.B.: *Fragile Brilliance: The Troubled Lives of Herman Melville, Edgar Allan Poe, Emily Dickinson and Other Great Authors.* Pythagoras Press, 2021, pp. 87-104.

9. Mendelson, W.B.: The Psychoanalyst and the Nazi Nobelist: *The Curious Story of Sigmund Freud and Julius Wagner-Jauregg*. Pythagoras Press, 2022.

10. Meynert, T.: *Psychiatrie. Klinik der Erkrankungen des Vorderhirns*. Intank Publishing, 2020 (original 1884).

11. Peled, A.: Neuroanalysis: the future of psychoanalysis. Heruka, June 17, 2019. https://herukahealthinnovations.com/2019/06/17/neuroanalysis-the-future-of-psychoanalysis/

12. Slatin, S.: Opera and revolution: La Muette de Portici and the Belgian revolution of 1830 revisited. J. Musicological Res. 3(1-2), 1979.

13. Stockert-Meyner, D.: *Theodor Meynert und Seine Zeit*. Austrian Federal Publishing House for Education, Science and Art, Vienna and Leipzig, 1930.

14. Wagner-Jauregg, J.: Lebenserinnerungen (L. Schönbauer and M. Jantsch, eds.) Venna, Springer, p. 72.

15. Whitrow, M.: Freud and Wagner-Jauregg: A historiographical study. Psychiatric Bulletin 14: 356-358, 1990.

Chapter Five:
Vulnerable Mentors
and Rebellious Students

Mentoring in the context of an evolving science:

Müller and Meynert were innovators who profoundly affected the trajectory of 19th century science; on the other hand, in their later years they may have remained focused on their early achievements, to the consternation of some of their students. In Müller's case, Emil du Bois-Reymond was critical of his mentor's seeming lack of interest in pursuing experimental physiology after 1840, while his students were using the emerging developments in physics and chemistry to make new discoveries. While he personally turned his attention back to his origins in comparative anatomy, and apparently realized that he would not be entering this new world himself, he clearly encouraged his students to do so. It might be that even though he realized that the results might not fit well with his own work, his curiosity about physiological processes was so great that he was willing to do this.

In contrast to Müller, Meynert, continued to concentrate on his original hierarchical and primarily anatomical view of the causes of mental illness.

While this had been revolutionary at the time, it left little room, for instance, for his student Carl Wernicke's later development of the notion of pathology resulting from disturbances in neural associative networks. Wernicke dealt with these differences graciously, while Sigmund Freud was quick to challenge the limitations of Meynert's anatomical views, favoring instead his conception that life experiences could result in functional disturbances in the nervous system. Meynert was to hold to his lesion-centric viewpoint for the rest of his career.

It's also important to remember that, whatever their shortcomings in their later careers, Müller and Meynert themselves had in their earlier years been innovators, participants in discarding *Naturphilosophie*, the then-dominant philosophical view of living organisms (see the Introduction), and replacing it with another approach, that is, experimental physiology. Müller had been brought up in an academic atmosphere dominated by *Naturphilosophie*. Upon coming to Berlin in 1823 it must have been a remarkable experience for him to study under Karl Asmund Rudolphi, who passionately rejected the Romantic movement and advocated empirical research. As we saw in Chapter One, by 1840 Müller had not fully reconciled the two, arguing that processes of living organisms could not be fully explained by a mechanistic approach, while also advocating the pursuit of empirical observation and experimentation "as far as we can". Certainly he was opposed to viewing physiology as a body of knowledge driven primarily by philosophical stances; on the other hand he felt that pure observation without any philosophical structure was fruitless. It is also hard to determine to what degree he maintained this view in his older years, as some of the students who later wrote his history were invested in portraying him as a vitalist to the end.

Meynert, in a later generation than Müller, was much more forceful in his rejection of *Naturphilosophie*. In explaining the scope of his 1884 textbook, for instance, he wrote that psychiatry, as he understood it, did not involve the historical concept of treating the soul, but rather encompassed the study of mental illnesses as a result of diseases of the forebrain (Meynert, 2020 edition). Though incorrect in the particulars, by emphasizing brain pathology he anticipated the development of modern biological psychiatry.

Personal qualities of the mentors: Another aspect shared by Müller and Meynert was a kind of social awkwardness, which particularly came out in dealing with their trainees. In Chapter Four, we described the efforts of Bernard Sachs to engage Meynert in small talk about the weather, only to be rebuffed, but also noted that his students generally felt that he tried to be amiable, but perhaps not cordial. When first getting to know Müller, du Bois-Reymond commented that "...to be friendly, he starts making the strangest faces' (du Bois Reymond, 1839). One can picture Müller, to whom good eye contact and a smile may not have come naturally, making an unsuccessful effort to put his new student at ease.

Müller and Meynert also shared a penchant for working in disorderly facilities. We have mentioned earlier how du Bois-Remond compared Müller's dissecting area to foul-smelling caves, while Auguste-Henri Forel likened Meynert's lab to Vienna's Oriental Quarter. As it happened, both students came from well-to-do families, and it is possible that their standards of comfort collided with those of Müller, the shoemaker's son, or of Meynert, whose Bohemian background emphasized creativity over orderliness.

Both mentors, as well, met their ends at relatively young ages (Meynert at 58, Müller at 56) as rather sad, lonely figures. Meynert grieved for the loss of his first wife Johanna in 1879 and his son Karl in 1884. He had settled in Klosterneuburg, to be close to their graves, and perhaps to take comfort in happy memories of the Weidlinger Valley. In 1892, the sickness and death of his daughter Christi no doubt weighted heavily on him when he passed away a few months later. In contrast, the major disruptions of Müller's life were internal, periods of mental instability leading up to the months in 1858 in which he suffered insomnia and palpitations, wandering the back streets of Berlin in the middle of the night, before dying from what was most likely an overdose of morphine.

Mentoring and emotional distress: Müller had at least three episodes of mental distress in his lifetime: in 1827 shortly before he was due to get married, in 1848 following the stresses he experienced as rector of the university during this period of political upheavals, and in 1858 preceding his death. These episodes have generally been considered as representing clinical depression. Although the retrospective diagnosis of historical figures is at best a hazardous undertaking, it is easy to understand how this conclusion was often reached, particularly regarding the period before his death. Sometimes overlooked, however, is his first episode in April 1827 (Chapter One), which may have been somewhat different. During this time, in addition to feeling irritable and unable to work, he believed that his legs were paralyzed due to a spinal cord lesion, and that he might be dying. As far as we can tell from Horst Müller's biography, however, thoughts of death were not suicidal, but rather a source of great anxiety. As noted earlier, there is no mention of feelings of hopelessness or helplessness, which often accompany depressive disorders.

Certainly in those years, physicians were well aware of diagnoses of melancholia. Robert Burton's *The Anatomy of Melancholy*, written some two centuries earlier, was well known in medical and literary circles; it was, for instance, said to be the favorite book of physician/poet John Keats, who was writing in the decades before Müller's episode. Nonetheless, Philipp von Walther, a prominent physician who had also been Müller's teacher in Bonn, reported to the state minister that he suffered from hypochondriasis. Though the meaning of hypochondriasis has shifted since being described by Thomas Sydenham in the 17th century, and also may have historically been used more often in men as opposed to a diagnosis of hysteria in women (Veith, 1965), it at least made clear that Müller's physician believed that no traditionally medical cause was associated with his symptoms. It is difficult to determine exactly what von Walther had in mind, however, because at that time the term 'hypochondriasis' was sometimes used in referring to depressive symptoms as well (Haberling, 1924). Since 'functional neurological disorder', the modern term encompassing paralyses without objective findings, is often accompanied by depressive disorders (Carson et al., 2000), it is also certainly possible that both were present.

Müller's episode including paralyses is interesting in that he was not the only prominent biologist of the time to have similar experiences. When in 1859, the year after Müller's death, Charles Darwin's *On the Origin of Species* was published, Darwin was residing at a remote heath spa in Yorkshire, in an effort to recover his health (Mendelson, 2021). In the following years, he consulted physicians for his many symptoms, which included crying spells, 'lumbago', 'rheumatism', and episodes in which he appeared to be unable to speak or to be partially paralyzed. His many prominent physicians were never able to find objective correlates of his symptoms until they recognized heart disease when he was in his 70s. In the interim, Darwin received a number of diagnoses, among which was hypochondriasis (Mendelson, 2021).

Ironically, one could speculate that among other 19th century scientific luminaries with unexplained medical symptoms was Theodor Meynert. In *The Interpretation of Dreams,* Freud reported that during his visit when Meynert was terminally ill, Meynert had commented that he himself had suffered from male hysteria. In Chapter Four, we covered a number of possible ways this could be understood; among them was the possibility that Meynert had indeed experienced unexplained symptoms, which had not come to medical attention. There is not enough information available, and it seems unlikely that we will ever know, but it is ironic that although Müller is remembered for his psychiatric symptoms, Meynert too may have suffered, albeit in a quieter, more private manner. They, along with Darwin, were among a group of nineteenth century biologists who made remarkable contributions even in the context of complex personalities and possible mental illness. Another example would be Matthias Jakob Schleiden, the botanist who had also been a student of Müller's, and who, along with Theodor Schwann, formulated the notion that all living things are comprised of cells, during a difficult lifetime which included two suicide attempts (Mendelson, 2024).

So far in this chapter, we have focused on the mentors. Let's turn now to some thoughts about the similarities and differences in their students— Sigmund Freud and Emil du Bois-Reymond, and the manner in which they reacted to their teachers.

The rebellious students:

An obvious difference between Freud and du Bois-Reymond is in their backgrounds—the son of a wool merchant, on the one hand, and on the other the privileged progeny from a family of scholars and artists. Conversely, it could be argued that a particular similarity played an important role in their lives: both were in some ways outsiders in their communities. Freud came from a Jewish family from the provinces who moved to largely Catholic Vienna, where he was very much aware of the widespread antisemitic atmosphere. He was also marginalized in a different way by Meynert, who emphasized that his training in hypnosis was by the French neurologist Jean-Martin Charcot, and mockingly referred to Freud as "a skilled practitioner of hypnosis back from Paris" (Chapter Four). Du Bois-Reymond, in turn, grew up in Berlin in a French-speaking family, and may have pronounced his German with a French accent. During the Franco-Prussian war, his French name was apparently an embarrassment to him. Though his cultured background may have opened some doors to him in polite Berlin society, there is also a sense that to some degree he was never fully accepted.

It is an interesting question whether this sense of not quite belonging influenced Freud and du Bois-Reymond's reactions toward their mentors. It is conceivable that it may have contributed to the vigor with which Freud reacted to Meynert's censure by setting up a private practice and conducting a career largely outside the traditional university setting (Mendelson, 2022). Freud may have dealt with a feeling of marginalization by creating his own community in which he was at the center of the stage—the psychoanalytic movement. Similarly, one might speculate that a sense of not being fully accepted contributed to the energy with which du Bois-Reymond argued that his mentor—Müller, who in the 1850s spent much of his time alone, cataloging the morphology of creatures in his museum—belonged to a past era while he, du Bois-Reymond, was in the mainstream of contemporary science.

Their reaction to their mentors: The overt responses of the two students to their mentors were also very different. As we have seen, du Bois-Reymond honored Müller with all the trappings of a loyal follower while he was alive; only after his death did he reframe the history of Müller's career in a way that was to his own advantage (Chapter Two). In contrast, Freud was very open in his disdain for some aspects of Meynert's science, criticizing him for his strictly anatomical view of mental symptoms, and his reluctance to accept the role of experiences in contributing to mental disorders. In Chapter Four we considered various ways of looking at Freud's claim in *The Interpretation of Dreams* that Meynert on his deathbed said that he himself had suffered from male hysteria. Another possible way of looking at it would be that Freud, in good faith with his recollection, had some similarities to du Bois-Reymond in that he was re-writing for history the narrative of his mentor's life.

The tables turned: Freud as a mentor: Twenty years later, when Freud in his 50s was a mentor to a host of brilliant students, he, like Meynert and Müller years before, was seen by some of them as being stuck at an outdated point and not being able to move on. Carl Jung, whom Freud had initially thought of as his possible successor, was critical of Freud for remaining focused on sexuality as the basis of neuroses. While Freud viewed the unconscious as being populated by sexual and aggressive urges, Jung believed that it also contained archetypes found in all human beings, and chided Freud for being unable to accept this much broader view of its function (Ellenberger, 1970). Similarly, Alfred Adler believed that Freud's focus on sexuality blinded him to the importance of social relationships and particularly to the 'will to power' (McCluskey, 2021). In both cases, Freud's response was overtly angry and sarcastic, with a dismissiveness reminiscent of Meynert's response to him when he returned from Paris and began to lecture on Charcot's views of hysteria and hypnosis. The tables had turned since Freud was 30, and the rebel of the 1880s, now in the 1910s, found himself to be the mentor, accused by his erstwhile students of being unable to join them in moving forward into what they saw as new and promising areas.

Concluding comment:

It has become clear that the successful role of teacher—and a medical teacher in particular—involves much more than just the imparting of information. It has been suggested, for instance, that a good teacher also serves as a role model, a facilitator, assessor, planner and resource developer (Harden and Crosby, 2000). The stories we have seen here—of Emil du Bois-Reymond and Johannes Müller, and of Sigmund Freud and Theodor Meynert— suggest that being a role model may be more complex than simply being a person to be emulated. To be sure, sometimes the dominant relationship is one of warmth and mutual respect, as was the situation with Meynert and Carl Wernicke. In others, notably with du Bois-Reymond and Müller as well as Freud and Meynert, from the student's point of view the teacher served as a figure who needed to be rebelled against in order for progress to be made.

Figure 5-1: *Interestingly, this theme of a student assailing his mentor has roots in literature at least as far back as ancient Greece in the story of Chiron and Heracles. The centaur Chiron, half-man and half-horse, who was wise in medicine and life in general, was mentor to many of the Greek heroes, including Jason, Achilles and Asclepius. He met his match, however, in his student Heracles, who incidentally had a history of previously killing his music teacher after being*

criticized. Later, during a fight with some other centaurs, Heracles is said to have accidentally wounded Chiron, whom he respected, with an arrow dipped in the serpent Hydra's blood. The pain was too great even for the demi-god healer, and he later gave up his immortality to ease his own suffering. Thus Heracles ended Chiron's life as a teacher, though in another sense he lived on: Zeus, recognizing Chiron's greatness, found him a home among the stars where he still resides as the constellation Centaurus. This 17th century depiction of Chiron and Achilles is found in Brussels in the Royal Museums of Art and History.

There are a number of ways to interpret the source of the rebellious stance against the teacher. Otis (2007) suggests that this may sometimes involve the projection onto the teacher of the student's relationship to the father: du Bois-Reymond is said to have treated both his father and Müller similarly, picturing both as advocates of outmoded ways of thinking who nonetheless are pleased with the son's/student's success. The manner in which the teacher is painted may also change at different points in the student's career, as in the case of Ernst Haeckel (Chapter Two), who, depending on his personal needs of the moment, attributed to Müller the roles of outdated proponent of comparative anatomy or of the earlier innovative experimental physiologist (Otis, 2007).

In summary, Müller and Meynert were remarkable scholars and teachers who left an indelible mark on scientific thought in the early and late 19th century respectively. Müller is known for grounding physiology in careful observation and experimentation, as well as bringing together all the contemporary knowledge of biological systems. Meynert, a generation younger, is credited with formulating psychiatry as a scientific discipline with its roots in brain pathology. Both are known as well for bringing together a distinguished group of students who went on to create disciplines including experimental electrophysiology, cellular pathology, and physiological optics.

Though many of the relationships of Müller and Meynert with their students were very positive and tolerant of intellectual differences, in the two cases we have examined here, the later achievements of the students were built on steadfast rebellion. Both mentors faced this insurgency while also coping with their own internal struggles and difficult events in their lives. On a more personal level, both spent their final years in apparent sadness and isolation. And yet somehow, out of all this internal and external turbulence came remarkable achievements in 19th century thought.

REFERENCES FOR CHAPTER FIVE

1. Carson, A.J. et al.: Do medically unexplained symptoms matter? A prospective cohort study of 300 new referrals to neurology outpatient clinics. J. Neurol. Neurosurg. Psychiatry. 68(2): 207-210, 2000.

2. Du Bois-Reymond, E.: letter, December 7, 1839 (cited by Otis, 2007).

3. Ellenburger, H.F.: *The Discovery of the Unconscious: The History and Evolution of Dynamic Psychiatry.* Princeton University Press, 1970.

4. Freud, S.: *The Interpretation of Dreams,* 1900. In James Strachey's translation of Volumes IV and V of the 1953 *Standard Edition* (London, Hogarth Press and the Institute of Psychoanalysis), Basic Books, New York, pp. 527-528.

5. Haberling, W.: Johannes Müller: Das Leben des Rheinischen Naturforschers. Leipzig, Akademische Verlagsgesellshaft, 1924, 2024.

6. Hagner, M. and Wahrig-Schmidt, B.: *Johannes Müller und die Philosophie.* Berlin, Bakeemie Verlag, 1992.

7. Harden, R.M. and Crosby, J.: AMEE Guide No 20: The good teacher is more than a lecturer—the twelve roles of the teacher. Medical Teacher 22(4):334-347, 2000.

8. McCluskey, M.: Revitalizing Alfred Adler: an echo for equality. Clin. Social Work 50(4): 387-399, 2021.

9. Mendelson, W.B.: *Fragile Brilliance: The Troubled Lives of Herman Melville, Edgar Allan Poe, Emily Dickinson and Other Great Authors*. Pythagoras Press, 2021, pp. 87-104.

10. Mendelson, W.B.: *From Despair to Discovery: The Botanical Odyssey of Matthias Jakob Schleiden and the Dawn of Cell Theory*. Pythagoras Press, 2024.

11. Mendelson, W.B.: *The Psychoanalyst and the Nazi Nobelist: The Curious Story of Sigmund Freud and Julius Wagner-Jauregg*. Pythagoras Press, 2022.

12. Meynert, T.: *Psychiatrie. Klinik der Erkrankungen des Vorderhirns*. Intank Publishing, 2020 (original 1884).

13. Müller, J.: *Ueber die phantastischen Gesichtserscheinungen*. Koblenz, Jacob Hölscher, 1926.

14. Veith, I.: Hysteria: *The history of a disease*. The University of Chicago Press, 1965.

Picture Credits

The author's assessment is that all images are in the public domain or presented under the terms of Section 107 of the U.S. Copyright Law (the 'Fair Use' provision). When appropriate, all reasonable efforts have been employed to trace copyright holders and to get their permission for the use of copyright material. The author apologizes for any errors or omissions in this list and will gratefully include any corrections in future editions if notified.

The author and Pythagoras Press do not have responsibility for the persistence or accuracy of URLs for external or third-party internet websites referred to in this publication and do not guarantee that any content on such websites is, or will continue to be, accurate or appropriate.

Cover Illustration: (Charcot) Etching by A. Lurat (1888), after P.A.A. Brouillet (1887). Attribution 4.0 International (CC BY 4.0), The Wellcome Collection/Public domain.

Figure 1-1: (Müller histology of various types of tumors). Muller, Johannes: *Ueber den feinern Bau und die Formen der krankhaften Geschwülste. Erste Lieferung,* Berlin : G. Reimer, 1838. From Wellcome Collection, Public Domain.

Figure 1-2: (Virchow histology): Virchow, R., Die Cellularpathologie : in ihrer Begründung auf physiologische und pathologische Gewebelehre zwanzig Vorlesungen gehalten während der Monate Februar, März und April 1858 im pathologischen Institute zu Berlin. Wellcome Collection/Public Domain.

Figure 2-1: (horse electrocuted by eels): Close-up of title page from From Untersuchungen über thierische Elektricität / Von Emil du Bois-Reymond / Volume 2 / 1849. Wellcome Collection/Public domain.. https://wellcomecollection.org/works/n6sagcfe

Figure 2-2: (von Helmholtz pendulum). Von Helmholtz' pendulum. For measuring the speed of the nerve impulse..In Exhibition 1972-3. Wellcome Collection/Public domain. Attribution 4.0 International (CC BY 4.0)

Figure 3-1: (medial surface of the human brain) Meynert, T.: Psychiatry: A clinical treatise on diseases of the forebrain based upon a study of its structure, Functions and Nutrition (translated by B. Sachs), G.P. Putnam's Sons, New York, 1885. From Wikimedia Commons/Public domain.

Figure 4-1: (Freud's microtome). A brain-tissue slicing instrument inside Sigmund Freud Museum in Vienna. Slyronit in Wikimedia Commons/Public Domain.

Figure 5-1: (Chiron) *De centaur Chiron onderricht Achilles.* Attributed to Frans and Jan Raes (after a design by Peter Paul Rubens), Wikimedia Commons/Public Domain.

www.ingramcontent.com/pod-product-compliance
Lightning Source LLC
Chambersburg PA
CBHW050548280326
41933CB00011B/1769